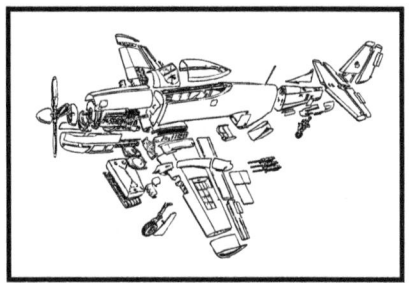

VOLUME 39

LOCKHEED C-141
STARLIFTER

FREDERICK A. JOHNSEN

Published by
Specialty Press Publishers and Wholesalers
39966 Grand Avenue
North Branch, MN 55056
United States of America
(800) 895-4585 or (651) 277-1400
www.specialtypress.com

COPYRIGHT © 2005 FREDERICK A. JOHNSEN

ISBN 978-1-58007-210-6
Item # SP080P

All rights reserved. No part of this publication may be reproduced or utilized in any form or by any means, electronic or mechanical, including photocopying, recording, or by any information storage and retrieval system, without prior permission from the Publisher. All text, photographs, and artwork are the property of the Author unless otherwise noted or credited.

The information in this work is true and complete to the best of our knowledge. However, all information is presented without any guarantee on the part of the Author or Publisher, who also disclaim any liability incurred in connection with the use of the information and any implied warranties of merchantability or fitness for a particular purpose. Readers are responsible for taking suitable and appropriate safety measures when performing any of the operations or activities described in this work.

All trademarks, trade names, model names and numbers, and other product designations referred to herein are the property of their respective owners and are used solely for identification purposes. This work is a publication of CarTech, Inc., and has not been licensed, approved, sponsored, or endorsed by any other person or entity. The Publisher is not associated with any product, service, or vendor mentioned in this book, and does not endorse the products or services of any vendor mentioned in this book.

Printed in USA

Cover: *The seventh C-141A built (63-8076) came in for a landing at Dobbins Air Force Base, Georgia, on July 5, 1964, with grass and pavement reflected in its shiny natural metal skin. (Photo by Kenneth G. Johnsen)*
Title Page: *With the Starlifter's petal doors in the ground loading position (open wider than the airdrop mode), a U.S. Army UH-1 helicopter, minus rotor blades, is eased inside, circa August 1982.*
Top Right: *A camouflaged C-141B of the 62nd Military Airlift Wing executed a steep roll to the right during a mission over western Washington in April 1983.*
Top Left: *Flare dispensers for defeating heat-seeking weapons were added to some C-141s, some of which were also upgraded to C-141C models. (Air Force photo)*
Bottom Right: *The first of the breed, C-141A, construction number 6001 (Air Force serial 61-2775), as it appeared circa January 1963 during fuselage section mating. The main landing gear pod fairing is visible in the right of the photo. (Lockheed Martin)*

Table of Contents

Lockheed C-141 Starlifter

	Preface 4
Chapter 1	Origins of Airlift and the Rise to the C-141 . . 5
Chapter 2	Starlifter Described 13
Chapter 3	What Might Have Been and Look-Alikes 33
Chapter 4	Getting the Job Done 42
Chapter 5	Paint and Markings 52
Chapter 6	C-141 Development and Operations 60
Bonus Section	Starlizards and Silver Bullets 65
Chapter 7	Flying the Starlifter 86
Appendix A	C-141 Serial Numbers 97
Appendix B	Starlifters Preserved 99
Appendix C	C-141 Significant Dates 101
	References 103

PREFACE

AND THANKS TO THOSE WHO'VE HELPED

Lockheed's swept-wing C-141 enabled the U.S. Air Force to realize the promise of a global airlift that the Douglas C-124 foretold in the slower piston-engine era. With the Starlifter, cargo and troops could be airlifted intercontinentally at three-quarters the speed of sound. From the 1960s to the 21st century, C-141s have been a trusted transporter of everything from returned hostages to presidential limousines.

So efficient and reliable are C-141s that they have been an expected part of the orchestra of global events during four turbulent decades that saw the culmination of the Cold War and the rise of terrorism as a major military and civil concern. The Starlifter's story is overdue.

Some convention is in order when discussing the C-141. Lockheed christened it the StarLifter, but almost universally the Air Force and others have chosen to say it with a lower case "l" as Starlifter, and that is the form used in this narrative. The quest for numbers – dimensions, weights, and performance – can be a maddening hunt among authoritative tables that differ. Such was the case when researching the C-141. The reader will notice that, as the C-141 proved itself, both its gross weight and rated engine life increased. The word "Wing" with a capital "W" in the middle of a sentence will refer to the particular Air Force airlift wing elsewhere described in that portion of the narrative. And the nomenclature of Air Force units will reflect the name in use at the time under discussion – sometimes Air Transport Wing (ATW), Military Airlift Wing (MAW), or just Airlift Wing (AW).

This book was propelled by the assistance, and collections, of a number of people, both inside and outside the Starlifter community, including the Air Force Flight Test Center History Office's Jeannine Geiger and Freida Johnson, Air Mobility Command History Office (and Betty Kennedy and John Leland), Air Mobility Command Museum (and Mike Leister and Harry Heist), Tony Accurso, Boeing Co. historian Michael Lombardi, Shirley Glaze, Marty Isham, Carl M. Johnsen, Kenneth G. Johnsen, Sharon Lea Johnsen, Craig Kaston, Pete Lochow, Lockheed Martin (and Jeffrey P. Rhodes), Don Macaluso, Mrs. Stephen McElroy, Doug Nelson, Doug Remington, Travis Air Force Museum (and Gary Leiser), U.S. Air Force Museum (and Sherry Howard), Terry Vanden-Heuvel (AMARC Public Affairs), and Linda Young.

My association with the Starlifter began when my father, aeronautical engineer Carl M. Johnsen, moved the family to Marietta, Georgia, in early 1962 as he took a job with the C-141 team. As a 12-year-old, I watched the first Starlifter, 61-2775, roll out of the Lockheed-Georgia plant on August 22, 1963. Thirty-five years later, I flew on this same venerable test Starlifter as it made its final flight to the Air Mobility Command Museum at Dover Air Force Base, Delaware. Between those two bookend events, throughout most of the decade of the 1980s, it was my extreme good fortune to participate in global airlift missions aboard C-141s of the 62nd Military Airlift Wing (MAW) based at McChord Air Force Base, Tacoma, Washington. I remember the good-natured professionalism of the aircrews, maintenance troops, and public affairs officers of the 62nd MAW and its Reserve Associate Wing, the 446th MAW.

With ranks omitted (since people may have received subsequent promotions), and with apologies to anyone inadvertently overlooked, I want to acknowledge many remarkable C-141 airlifters including: Robert J. Boots, Gordon L. Boozer, Don Buescher, Frank Cardile, Carl Chop, Bill (Omar) Chramosta, Stu Farmer, Dave Feigert, Jerry P. Harmon, Edward G. Hoffman, Ben Howser, Howard J. Ingersoll, Vernon J. Kondra, Keith Laird, Keith Littlefield, Tim Louden, John T. Loughran, Michael J. McCarthy, Jerry McKimmey, Gary H. Mears, Herb Mellor, Richard C. Milnes, Donald C. Smith, Doug Strodbeck, Edwin E. Tenoso, and Hugh Wild.

Frederick A. Johnsen
2005

P.S. As this volume enters its second printing, the last Starlifter flying is ready for retirement with honors in May 2006, outlasting the Cold War in which it was conceived. Thousands of people expressed interest in the C-141 retirement ceremony. With a quiet nod and a faraway gaze, as long as airlift veterans gather, they will acknowledge the pioneering legacy of the Lockheed C-141 Starlifter. – FJ, 2006

Origins of Airlift and the Rise to the C-141

A rare in-flight convening of major strategic and tactical airlifters roared past simmering Mount St. Helens in October 1986 as the McChord Air Museum's newly acquired C-124C led a C-130E and C-141B from the 62nd Military Airlift Wing on the Globemaster's delivery flight to the museum. The C-124, long ago retired, was acquired from a vocational school in Michigan.

The piston-engine Douglas C-124C Globemaster II blossomed in the 1950s as an intercontinental airlifter, although its slow speeds and unpressurized altitude limitations hampered its utility in the dawning jet age. Low wing and large propeller arcs necessitated steep nose ramps to load vehicles, plus an aft ventral cargo elevator. This Globemaster II example flew to the McChord Air Museum in October 1986.

As civil and military aviation grew in the 1920s and 1930s, the mission of air transport remained modestly defined by commercial aircraft designs adapted in small numbers by the Air Corps. The technology-driven premise was that troops and materiel would be sent to war by ships; some localized air transport could be provided by general-aviation or airline-type aircraft.

In pre-World War II exercises, the Air Corps sometimes drafted bombers to carry supplies as makeshift transports. The war years saw extensive use of C-47 and C-53 variants of the twin-engine Douglas DC-3 airliner as an intra-theater transport, augmented by Curtiss C-46s that missed the call as first-line airliners, but served well as military transports. Transoceanic air transportation was provided by four-engine Douglas C-54s, many of which saw immediate postwar service as DC-4 airliners.

The wartime need for transports prompted bombers to be modified with varying degrees of success; the B-24 yielded the successful C-87 and C-109, while a handful of B-17s emerged as C-108 conversions. Yet ships remained the means for moving armies overseas. But the capacity of large airframes and powerful multi-row radial reciprocating engines was undeniable. As early as 1942, the service placed an order with Fairchild for the XC-82, a purpose-designed transport with a level deck and large cargo doors that could carry 42 troops or a variety of cargo options. With a listed range of 3,875 miles, the C-82 gave wings to the notion of intercontinental airlift, albeit on a modest scale. Meanwhile, Douglas used four 3,500-horsepower R-4360 engines to power its C-74 Globemaster of 1945, an early postwar contender. It was soon eclipsed by the follow-on C-124 Globemaster II of 1949 that started out with the C-74's wing, engines, and outsized tail surfaces and added a boxy double-deck fuselage with a fold-away floor that could be moved aside to accommodate large vehicles and cargo. The C-124 could carry 200 passengers. The ultimate C-model of the Globemaster II cruised at 230 miles an hour at 10,000 feet, and had a range of 4,000 miles while carrying more than 13 tons of cargo.

In 1948, the Soviet blockade of Berlin brought the strategic and humanitarian value of airlift to the fore as streams of C-54s, C-47s, and allied transports carried survival provisions into Berlin. It was a headline test of wills, and the airlift ultimately prevailed by the end of September 1949. In all, allied aircraft carried 2,343,301.5 million tons of supplies including foodstuffs and even coal into the city. American aircraft delivered substantially more than half of this total, at 1,783,826 tons.[1]

Undeniably, airlift capability was maturing in a Cold War arena where air transport would never again be viewed as simply in-country support. The C-124 Globemaster II gained a loyal following in airlift circles for its plodding reliability and prodigious capacity. Throughout the 1950s and well into the 1960s, the Air Force dispatched C-124s on both regular and non-scheduled airlift missions globally.

In April 1954, 13 Globemasters of the 62nd Troop Carrier Group flew from Washington state to France to pick up French troops. The troops were rapidly delivered to French Indochina in an ultimately failed bid by the French to stave off defeat at Dien Bien Phu. The C-124s made the globe-girdling trip, named Operation Bali Hai, in 8 to 10 days, with an itinerary that included Germany, France, Tunisia, Libya, Egypt, Saudi Arabia, Pakistan, Ceylon, Thailand, and Vietnam (French Indochina). The C-124s logged about 22,000 miles on this global flight, with an average of 119:07 flight hours. This feat would have been unthinkable during World War II.[2]

The 1950s forged the role of global airlift as a practical, if low-and-slow, tool in the American air arsenal, and created the stereotypical happy-wanderer airlifter, a can-do flier likely to know more about the neighborhoods he visited in foreign

Though essential C-141 design remained faithful to early Lockheed concept art, some changes occurred after this painting was executed. The Air Force eliminated an early requirement for a forward cargo loading door on the left side of the fuselage; the upper aft fuselage later acquired its characteristic hump, not evident in this picture; and the petal doors were designed to hinge along the upper seam. In the drawing, petal doors appear to have an earlier configuration that hinged along the forward vertical seam. Wind tunnel tests also subsequently dictated changes to the streamlined fairing at the junction of the vertical and horizontal stabilizers. Actual cockpit glazing also was different from the angle-cut rendition in this painting. Also, no overhead windows were used in the cockpit area. (Lockheed via Travis Air Force Museum collection)

lands than his own back yard. Developments in aeronautics set the stage for the next airlift generation – the fanjet-powered, pressurized, Mach .74-cruising C-141 Starlifter. If the C-47 and C-54 showed the promise of airlift in World War II, and the C-124 took it around the world in the 1950s, the jet age made airlift a real-time tool of Cold War diplomacy starting in the 1960s.

Flexible Response Helped Secure the C-141's Niche

As the Cold War verged into the early 1960s, some American military planners, realizing not all threats could be countered with a nuclear deterrent, espoused the theory of flexible response. Flexible response

An early color rendition of C-141As by a Lockheed artist incorporated bright bands of orange, in the years just before the Air Force returned to camouflage to conceal, rather than reveal, its aircraft. In service, C-141As in bare metal did not use bright visibility enhancing markings, although an Air Force paint specification was drawn to show where arctic orange high-visibility markings could be applied if needed. (Lockheed Martin)

Lt. Gen. William H. Tunner, MATS commander and an airlift visionary, set forth his wish list for a new-generation cargo aircraft, helping define the ultimate configuration of the Starlifter. The airframe needed to allow loading from standard truck-bed height, with straight-in access through the front or rear of the fuselage. This would necessitate the high-wing design. (Brig. Gen. Stephen McElroy collection)

The first Starlifter's wings were moved into position in the huge Air Force plant Lockheed operated in Marietta, Georgia. (Lockheed via Harry Heist, Air Mobility Command Museum)

called for the rapid deployment of military assets to trouble spots around the world. Airlift was key to the execution of flexible response, and old reliable transports like the piston-engine C-124 Globemaster II were too slow for the envisioned task.

Still, airlift's strongest proponents were airlift officers, and the concept of intercontinental high-speed purpose-built cargo aircraft did not enjoy universal support in the armed forces or in Congress. As described by Military Airlift Command (MAC) historians Roger D. Launius and Betty R. Kennedy, "airlift, in essence, did not neatly fit into the scheme for the optimal use of air power. It remained a stepchild – an auxiliary force – not contributing directly to the quest for air superiority or strategic bombardment."[3]

Two important agents of change in the status of American airlift were President John F. Kennedy and his Secretary of Defense, Robert S. McNamara. Though sometimes criticized

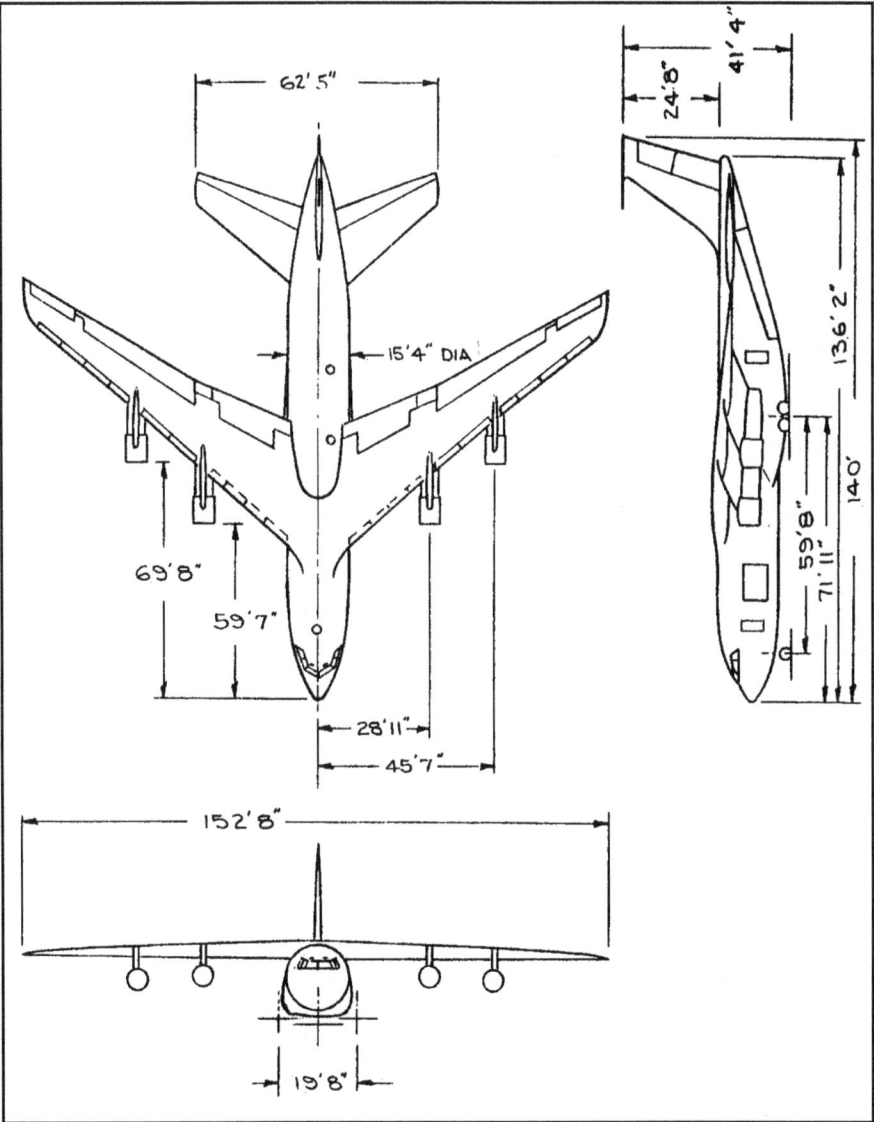

Boeing's entry in the C-141 contest was the company's Model 731, a conventional-tailed high-wing jet transport with a proposed wingspan of 152 feet, 8 inches and a length of 140 feet. (The Boeing Company)

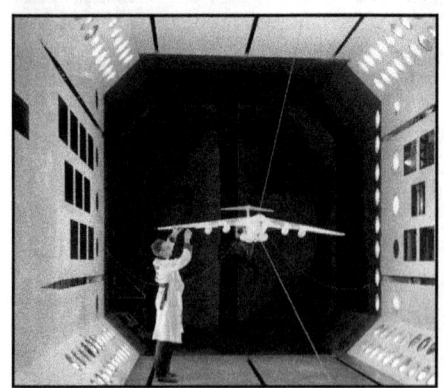

C-141A wind tunnel tests at NASA's Langley Research Center honed the T-tail design before production. (NASA Langley photo)

for actions he espoused, it is evident McNamara – and his President – elevated the need for a new airlift aircraft to make flexible response achievable. This heightened interest in the potential capability of airlift helped push aside any lingering reticence on the part of some in the military and Congress, making the proposed new jet airlift airplane a priority that enhanced the very status of the Military Air Transport Service.[4] President Kennedy judged the C-141 to be important enough to warrant his personal attention during the first Starlifter's rollout August 22, 1963, when, via circuits between Washington, D.C. and Marietta, Georgia, the President pushed a button that sent a signal to open the Lockheed plant doors enabling the first C-141A to be rolled out in the bright Georgia sun.

Company artist depicted the simple streamlined future of airlift in a C-141A rendering. (Lockheed Martin

In such a climate, the genesis of the C-141 Starlifter brought American military airlift to the forefront. Later, Gen. Howell M. Estes, MAC commander when the C-141 entered the inventory, said: "Global military airlift has been shown, throughout the era of the Cold War, to be a principal medium of achieving maximum military flexibility."[5] Starlifter crews interpreted it succinctly: while fighter and bomber pilots practiced their missions, airlifters were busy executing theirs, day in and day out, all over the world.

By 1968, Starlifters in MAC service had proven capable of deploying troops in several events, including a Cold War standard, Reforger (a convenient contraction of Return of Forces to Germany). Statistics for fiscal year 1968, three years into C-141A operations, showed Starlifters carrying more than 82 percent of MAC's standard "channel" mission cargo, while C-141s ended that fiscal year representing only slightly more than 50 percent of the aircraft in the MAC transport force.[6]

The C-141 concept used generally proven components to create a new airplane that would spearhead a new role for airlift. The program was premised on concurrent development and production; there was no XC-141 prototype. The first aircraft built was a C-141A, essentially a production machine. Even as developmental flight testing was taking place at

The Starlifter's moderate 25-degree swept wing, evident in this overhead view of a C-141 and a straight-wing C-130 on the Lockheed-Georgia ramp. The 25-degree sweep was sufficient to allow speeds of Mach .74 and above, while retaining low-speed handling characteristics that did not require the use of complex movable leading-edge devices. (Lockheed Martin)

Lockheed's logistics line-up, circa 1968, included (left to right) the gargantuan C-5A Galaxy, the C-141A Starlifter, and the C-130 Hercules. Just as the Starlifter benefited from concepts proven by the C-130, the C-5 was the beneficiary of design ideas validated with the C-141. (Lockheed Martin)

Edwards Air Force Base, more C-141As were rolling off the Lockheed line. The result was the rapid introduction of jet Starlifters into the airlift mix, with initially limited operating regimes. Testing cleared other areas of the envelope, and the inevitable deficiencies uncovered in testing were corrected. This style of acquisition program was repeated with the Starlifter's ultimate replacement, the McDonnell Douglas (now Boeing) C-17 in the 1990s.

Lockheed Capitalized on C-130 Success

Lockheed was early to grasp the utility of high-wing transport designs with low-slung fuselages and gaping openings at the rear of the cargo compartment, making drive-on and drive-off loading feasible. The useful C-130 Hercules propjet validated this layout as soon as the first example flew in August of 1954. Shunning the twin-engine, twin-boom philosophy Fairchild had employed with its smaller C-82 and C-119 transports, the Lockheed C-130 used a large airframe traditional in many respects, albeit equipped with pod-mounted landing gear attached to the fuselage, instead of long main gear legs as had been used on other high-wing aircraft like Convair's World War II B-24/C-87 series. In some ways, the C-130 looked like an enlargement on the design tenets of the smaller 1949 Chase (later Fairchild) C-123.

But the C-130 did more than previous low-slung transports; it cruised at more than 300 miles an hour, better than 100 miles an hour faster than the C-123B, and out carried the piston twin-engine C-123B by more than 35 tons.

In late 1950, Lockheed had been induced to re-open the slumbering Air Force Plant No. 6, the giant manufacturing facility in which Bell Aircraft had made B-29 Superfortresses in Marietta, Georgia during World War II. Lockheed initially received contracts to refurbish B-29s there, followed by a slice of new B-47 Stratojet production. Lockheed exploited the huge floor space at Marietta for production of C-130s after building two YC-130 prototypes at Burbank, California. The logic was in place, and experience was being gathered; the stage was set for a new airlifter even more revolutionary than the turboprop C-130.

Lockheed's T-tailed swept-wing jet C-141 capitalized on the functional cargo capabilities of the company's already-proven propjet C-130 Hercules. For a number of years, the 62nd Military Airlift Wing operated two squadrons of C-141s and one squadron of C-130Es, making it the only mixed Starlifter and Hercules Wing in the Military Airlift Command.

Circular fuselage cross section of the C-141, modified slightly by flat cockpit windows, was advantageous for pressurization. Landing gear pods were fairings; the main gear structure tied directly into fuselage members, and was not supported structurally by the bulging pods.

Setting the Stage

The Korean War, following on the heels of the Berlin Airlift, gave rise to discussions within the Air Force on the need for large cargo aircraft capable of inter-theater airlift as well as intra-theater missions. By the end of the 1950s, Army planners were embracing the concept of rapid response as a means of stifling incipient conflicts before an enemy could build up forces with traditional means of transport. The need for an aircraft like the C-141 had to be codified before the aircraft could be designed. As expressed by the C-141 design, the aircraft needed to be more than a vehicle for transporting troops quickly; it was to be a useful cargo carrier as well.

A few vocal members of Congress in the 1950s argued against giving the Military Air Transport Service (MATS) – the forerunner of the Military Airlift Command and later the Air Mobility Command – new transports. "Let the airlines carry the troops," they said. In this environment, Lt. Gen. William H. Tunner, MATS commander and an airlift visionary, set forth his wish list for a new-generation cargo aircraft. General Tunner wanted a big aircraft that could carry large items without dismantling them. The airframe should

Escorted by its replacement, a C-124C Globemaster II led a C-141B across Washington farmlands in October 1986 as the C-124 flew to the McChord Air Museum.

The first C-141A joined two previous icons of airlift, the C-130 and a Douglas C-47 left over from World War II, on the Lockheed-Georgia ramp in 1963. (Lockheed Martin)

The McChord Air Museum's C-124C led a C-130E and a C-141B past Mount Rainier in October 1986 as the Globemaster flew to its new museum home from a vocational school in Michigan. This side view shows C-124's deep forward fuselage to accommodate steep loading ramps to an opening nose section.

A view from above emphasizes main landing gear pods on a 62nd Military Airlift Wing C-141B flying away from Mount Rainier in April 1983.

This takeoff photo shows large Fowler flaps extended and dropped behind the wing of a C-141A to augment lift. As built, C-141As typically carried their Lockheed construction number (in this case, 6035) on the sides of the nose. This Starlifter was assigned Air Force serial number 64-0622, as noted by the digits 40622 on the vertical fin. (Lockheed Martin)

With gear retracted for flight, this C-141A photographed in September 1978 presented a simple, straightforward streamlining effort. Bulging main landing gear pods were a necessary concession to preserve open space inside the cargo compartment. (Photo by Frederick A. Johnsen)

allow loading from standard truckbed height, with straight-in access through the front or rear of the fuselage. This would necessitate a high-wing design. In the late 1950s, General Tunner envisioned an aircraft that could carry an "appreciable" load from California nonstop to the Far East, and a much heavier load as far as Hawaii. He did not place great emphasis on top speed other than to call for the use of turboprop or turbojet engines, since every use of this airplane for intercontinental delivery of military cargo previously sent by surface transportation would realize tremendous time savings already.

By early 1959, General Tunner's farsighted cargo airplane concept, the Army's evolving airlift needs, and the sometimes-contentious interest of Congress were coalescing. On February 16, 1959, MATS submitted to USAF headquarters a Qualitative Operational Requirement (QOR) for a Logistic Aircraft Support System. The QOR, signed by General Tunner, furthered his premise, while acknowledging even larger airframes would be needed to carry truly outsized cargo.

The Air Force was faced with budget constraints that threatened to make only a short run for a large transport being touted in some circles. MATS and Air Force headquarters seized on the medium size of the aircraft envisioned in the QOR as a viable alternative. General Tunner said the service needed both the large and the medium airlift capacity, but in the interim, Douglas C-133 turboprops could carry the outsize items while the newly envisioned medium airlifter became the backbone of Air Force transport. Throughout 1959, the Air Force honed its requirements for a new medium airlifter. The demonstration of existing airlift capability in Operation Big Slam/Puerto Pine in March 1960 showed the inadequacies of the current system and helped gain favor in Congress for airlift modernization.

During congressional hearings, the Air Force presented its rationale for a new airlifter, saying expected limits on funding dictated the pursuit of only one new airlifter type at the time. Maj. Gen. Bruce K. Holloway, Air Force Director of Operational Requirements, told members of Congress: "…every effort has been made to evolve aircraft specifications that would meet the military requirements and, additionally, assure an aircraft possessing a high degree of commercial compatibility." Given the obsolescence of transports then in use, the new requirement called for a state-of-the-art airlifter instead of a long-lead design, to make the new airlifter available more quickly. The general said the new aircraft's design "must encompass performance and reliability features which will not only assure its effective utilization under widely varying conditions but also assure the continued effectiveness of this aircraft in order to preclude early obsolescence." He argued that only the development of a new-design optimum cargo aircraft would fill the bill; off-the-shelf procurement of existing cargo or commercial jet transports just wouldn't work.[7]

It was a time when Army officials suggested the C-130 could do the job, and FAA administrator Elwood Quesada had even temporarily expressed interest in the Canadair CL-44 turboprop freighter, a traditional low-wing design that did not address some of General Tunner's key points. Ultimately, Congressional queries led to the agreement to support the Air Force in its quest for a new optimum airlifter. By November 23, 1960, with Congressional support, the Air Force issued Development Directive 415 for the new airlifter, specifying the use of turbofan engines. In that election year, both major presidential candidates expressed support for what would later become the C-141.

As the Starlifter was taking shape in 1961, a memo circulated among MATS staff reminding them: "This is the first opportunity afforded MATS to participate in the entire development of a Strategic Airlift System." The opportunity was not to be wasted; MATS would have the world's best airlift aircraft by mid-decade.[8]

Both inboard thrust reversers appear deployed in this telephoto view of the number one Starlifter. The straight line of the taxiway provides reference for wing droop. (Lockheed Martin)

For show at Lockheed-Georgia, Army troops lined up beside a new C-141A, which was configured to carry 154 troops for airland delivery or 123 paratroops for airdrop. (Lockheed Martin)

STARLIFTER 2 DESCRIBED

Dimensions and Performance

All Starlifters were built at Lockheed's Marietta, Georgia plant as C-141A model equivalents, including one commercial L-300A demonstrator. In the late 1970s and into the early 1980s, the Air Force contracted Lockheed to stretch its operational A-models by 23 feet. The subsequent upgrading of small numbers of B-models as the C-141C did not change the dimensions from B-model data.

Selected Starlifter statistics are:

Model	Length	Span
C-141A	145'	160'
C-141B/C	168'4"	160'

Model	Height	Weight (Empty)
C-141A	39'3"	136,000 lbs
C-141B/C	39'3"	148,120 lbs

Model	Weight (Gross)
C-141A	325,000 lbs
C-141B/C	325,000 lbs*

*Emergency War Planning weight of 344,900 lb. ramp gross weight permissible with increased tire pressure and main landing gear strut pressure.

Model	Cruising Speed	Service Ceiling
C-141A	506 mph	41,600'
C-141B/C	490 mph	41,000'

Model	Range
C-141A	4,080 st. mi.
C-141B/C	2,575 nm*

*Can be extended with aerial refueling; assumes heavier cargo weight than A-model.

Although the C-141 can (with some peril) approach Mach .89, airlift crews were told never to exceed .825 Mach. In operational airlift use, the C-141 cruises at .74 Mach, or nearly three-fourths the speed of sound. The actual speed represented by .74 Mach varies with altitude. On

Forward fuselage of the first C-141A, construction number 6001, Air Force serial 61-2775, moved by crane in the Lockheed-Georgia plant as the aircraft was assembled. The process of joining the forward, center, and aft fuselage sections of this Starlifter began on January 11, 1963. (Lockheed via Air Mobility Command Museum)

The famous Hanoi Taxi *soldiered on long enough to become a stretched C-141B later upgraded to C-model status with a glass cockpit instrumentation suite, as photographed in July 2004. Separate throttles for pilot and copilot are visible in this view.*

The fan duct cowl sections can be opened to provide access to the C-141's TF33 engines. (Air Force photo via Col. Donald Macaluso)

Small rectangular panels beneath the wing of a C-141C mark the location of a defensive system flare ejector. Flares are carried in the wings and in a ventral fuselage location on modified Starlifters to counter heat-seeking missiles. Additional flare dispensers can be located in the main gear pods if needed.

Flare dispensers for defeating heat-seeking weapons were added to some C-141s, some of which were also upgraded to C-141C models. (Air Force photo)

certain occasions, if urgency dictates, pilots set the throttles to cause the C-141 to fly faster than .74 Mach; they call it "hurry home power." But the norm for the Starlifter is .74 Mach, and exceeding this speed usually requires justification.

The C-141A was initially designed for a maximum flight weight of 316,100 pounds (although higher gross weights at time of engine start were listed). It could land at this high weight as long as sink rate was kept no higher than six feet per second.[9] This would be especially important in take-off emergencies requiring an immediate return to base. In normal service, the C-141 could be expected to land with its heavy cargo still aboard, even though fuel consumption lowers the gross weight by mission's end. The purpose-built C-141 was built stout enough to land with its cargo onboard; some old World War II transport conversions like the C-87 Liberator Express variant of the tricycle-gear B-24 bomber experienced nosewheel failures when landing with cargo aboard. This was a problem not normally encountered in bomber B-24 variants, which typically landed at much lighter weights after dropping their bombs.

As Starlifters matured in the airlift system, higher weights were recorded in documents. Many sources list gross weight for the C-141A and B at 325,000 pounds, later increased to a maximum ramp gross weight of 344,900 pounds. The 344,900-pound ramp weight included a ton of fuel that was expected to be consumed in taxiing before takeoff.

Construction Traits

The C-141A was conceived as an amalgam of existing technologies; it was to be state-of-the-art – not advance it. The Air Force made a conscious choice to limit engineering changes during design of the Starlifter. Where other airlift aircraft had undergone more than 150 engineering design changes before rollout of the first example, the Starlifter program generated only 50 proposals for design changes, more than half of which were not approved.[10]

The 25-degree swept wing, while giving the Starlifter a speed advantage, does not employ any leading edge slats or leading-edge airfoil augmentation. In fact, this gentle degree of sweep contributes to low-

Both sides of the vertical fins of modified Starlifters carry protruding infrared sensors to scan the aft hemisphere for signs of missiles fired at the C-141 when near the ground. The only white-top C-141 still serving in the 21st century, this aircraft is the historic Hanoi Taxi (66-0177), given a retro paint scheme to commemorate its service as the first C-141 carrying POWs from Hanoi.

The C-141C Hanoi Taxi displayed at the EAA AirVenture show in Oshkosh, Wisconsin in July 2004 shows placement of an infrared detection sensor on the left side of the forward fuselage. Forward crew door rolls up and in to clear the opening.

speed flight controllability without the need for more sophisticated flap or slat employment in the C-141. The wings are attached to a pair of forged fuselage rings rather than directly to the center carry-through airfoil section. Conventional fowler-type flaps roll out and down behind the 160-foot wing, increasing its no-flaps area of 3,228 square feet in the process. Two distinct sets of flaps per wing accommodate differences in the trailing edge sweep, which becomes more pronounced with the outboard set of flaps. On landing rollout, the flaps can help the brakes and thrust reversers diminish the ground roll.

During Starlifter design, consideration was given to the unwanted Dutch roll phenomenon, a combined rolling and yawing motion. The C-141's built-in negative dihedral of 1.2 degrees, which put the wingtips 42 inches below the wing root, was intended to thwart Dutch roll. The outer wings flex, and in flight the C-141's wingtips are about 20 inches higher than at rest. A three-degree wing washout twist modifies the harshness of stalling characteristics. Blown hot air used for de-icing

The infrared sensor visible on the right forward fuselage of C-141C 66-0177 in July 2004 is one of four sensors arrayed to give maximum coverage to detect the approach of missiles from the ground. The sensors and flare dispensers provide protection during operations close to the ground, where missile attacks could take place.

the wings exits the wings through vents under the wingtips.[11]

A circular fuselage cross section (essentially the same diameter as that of the C-130) used for the C-141 makes some pressurization issues simpler to deal with than in an asymmetrical shape, but the Starlifter's huge hinged reinforced pressure bulkhead at the aft of the cargo compartment proved to be a vexing problem until several latch fixes were tried. The C-130 Hercules was so

seminal in cementing Lockheed's role as a builder of military transport aircraft that the future Starlifter was briefly called the Super Hercules before the aircraft was built.

Most overhead hatches, entries, and exits in production Starlifters open in, or are removable to, the inside of the fuselage. This is a simple feature to minimize the likelihood of a decompression blowout at altitude, as the hatches and doors are physically restrained by the airframe around them. An exception is the forward dorsal hatch in the cargo compartment. Unlike two other dorsal hatches, this one opens out and is designed to be jettisonable in flight if needed. This hatch is not routinely opened on the ground; the inward-opening hatch on the flight deck is regularly opened to permit walkway access to the top of the fuselage, as well as to allow fresh air to enter the sometimes-stifling cockpit until the aircraft gets underway.[12] The outward-opening hatch was also replaceable with a modified hatch carrying a satellite communications (SATCOM) antenna when such devices became available.

The high-mounted flexible wing of the C-141 dictates the placement of all landing gear in the fuselage. Early studies showing wing-mounted landing gear to support some of the weight of engines and fuel were discarded; the length of a landing gear strut from the wing was excessive. Short fuselage-mounted landing gear represented a weight savings, too. Unlike the sleek B-47 Stratojet bomber, whose tandem mainwheels retracted flush inside the tapering cylindrical fuselage, the Starlifter uses main landing gear pods attached to the outside of the fuselage. The main landing gear pods provide aerodynamic benefits, but not structural support for the gear. Dual nosewheels retract into a well beneath the flight deck. The four-wheel main landing gear bogeys have a tread (centered on the outboard tires) of 20 feet, 2 inches; wheelbase (on the C-141A) is 53.0 feet.

The Starlifter's high T-tail keeps the horizontal tail surfaces well above any possible issues with loading vehicles or facilities. It incorporates a movable horizontal stabilizer for trimming and conventional elevators for pitch control. The horizontal stabilizer was built of smaller sections

Classic eight-position views of the second C-141A (61-2776) were taken with the Starlifter parked on the hard bed of Rogers Dry Lake at Edwards Air Force Base on June 11, 1965. Checkerboard markings on nose and tail are targets for theodolites used for measurements during flight testing. Three decades after this series of documentary photos was made, C-141 2776 was at Edwards again as part of the 418th Flight Test Squadron, testing an aileron control modification under the nickname Electric Starlifter *in 1996. (AFFTC History Office)*

Lockheed in-flight photography of the first Starlifter under test shows natural metal fuselage contours. It is evident that the wing flaps suffer some exhaust smudging when they're extended down behind engines. (Lockheed Martin)

permanently joined to make one unit. Hydraulics move the control surfaces. Front, center, and rear spars support milled skin panels to form the core of the vertical fin. A false aft spar is the attach point for the rudder; the fin's leading edge is built-up aluminum construction. Two passageways with lighting and ladders provide access to pitch trim and elevator hardware; the forward passage also gains access to the top of the horizontal stabilizer. Where moisture is prevalent, such as at McChord Air Force Base in Washington, C-141 crew chiefs learned to anticipate a drenching in water and any other pooled fluids when accessing the horizontal stabilizer hatch from inside the fin.

When preflight wind tunnel tests indicated flutter problems emanating from the fairing at the junction of the horizontal and vertical stabilizers, a larger hourglass-shaped fairing replaced the original bullet design, eliminating the problem, but at a weight penalty of about 40 pounds.

Roll control is performed with traditional ailerons at the trailing edge of the C-141's wingtips. The modest sweepback of the wing obviated any need for a second inboard set of ailerons as was done on some jets of greater sweep, where outboard ailerons could lead to wing-warping.[13] Nonetheless, aileron reversal at higher-than-normal speeds is possible in the Starlifter. Spoilers on the tops of the wings can be deployed to kill lift during landing rollout, as well as in flight to promote quick descent.

To accommodate cargo loading and aerial delivery, the C-141 uses two petal doors that are hinged at the top where they attach to the aft fuselage. A sequence of events opens the airplane's pressure bulkhead and the petal doors while lowering part of the aft fuselage to create a flat deck for airdropping palletized or containerized (Container Delivery System, or CDS) cargo. In mock-up form, the petal doors originally hinged at the front instead of the top. During design, the aft fuselage acquired its pronounced hump, part of the redesign of the cargo door and ramp assemblies. The aft loading opening provided by the opened petal

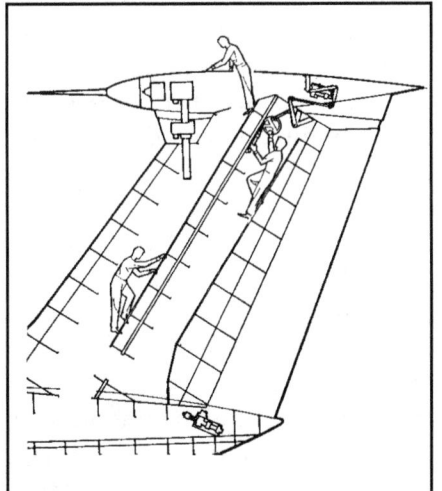

This line drawing shows access available inside C-141 vertical fin assembly, and provides a human scale reference. (Lockheed)

The C-141A was built to land within 3,700 feet after clearing a 50-foot obstacle, making modest length runways suitable for operations. (Lockheed Martin)

C-141A 63-8075, the first Starlifter with a 63- serial number, participated in flight tests at Edwards Air Force Base in California's Mojave Desert when photographed on June 15, 1964. The vertical fin and part of the petal doors were painted. (AFFTC History Office)

The C-141's engine nacelles culminated in the largest thrust reversers manufactured up to that time by the Rohr Corporation in Chula Vista, near San Diego, California. Left and right halves of the thrust reversers swung behind the engine exhaust when deployed on landing to help slow the Starlifter. The thrust reversers had outside skins constructed of aluminum. Interior surfaces, facing the hot exhaust, were made of heat-resistant A-286 alloy. Rohr also designed other C-141 cowling pieces to enclose the Starlifter's TF33 engines. (Air Force photo via Travis Air Force Museum)

doors is the same as the cross sectional dimensions of the cargo compartment: 10.25 feet wide by 9.1 feet high. The cargo deck is 50 inches above ground level; a hinged lower loading door can be positioned level with the rest of the deck for horizontal loading from vehicles of that height, or it can be lowered to meet the ground, with auxiliary ramps enabling drive-on, drive-off vehicle loading.

Unlike the slower turboprop C-130, the C-141 had a smoothly faired lower aft fuselage incorporating the petal doors. This reduced drag and contributed to efficiency. After the advent of the C-141, Lockheed gave consideration to incorporating a streamlined aft body on the workhorse tactical C-130 as well, but this was not implemented.

In the aft fuselage, two roll-up doors, one on either side, can accommodate paratroops exiting the C-141 on static lines. Starlifter fuselages are routinely fitted with red nylon web troop seating in rows with the occupants facing in toward the fuselage centerline. So-called "Boeing seats," built like traditional airliner seats, can be quickly mounted to the floor of the C-141 if desired. These airliner seats are mounted facing aft to provide some deceleration protection for passengers in the event of a mishap. For long flights with passengers, a comfort pallet featuring hot and cold kitchen facilities and an extra lavatory can be installed near the front of the cargo compartment.

  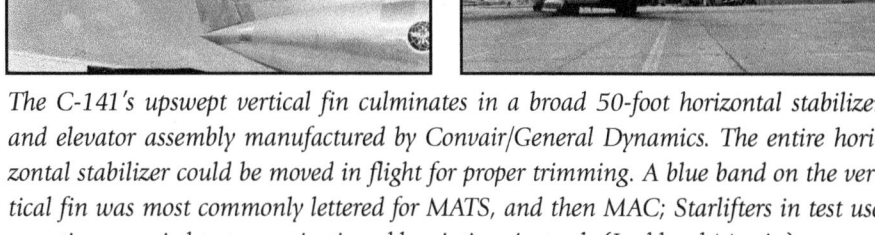

Two of the South's commercial successes — Lockheed's C-141 Starlifter and Coca Cola – happened to share the frame in a Lockheed photographer's view of the first Starlifter, coming over the perimeter at Dobbins Air Force Base in Marietta, Georgia, where the Lockheed plant was situated. (Lockheed Martin)

The C-141's upswept vertical fin culminates in a broad 50-foot horizontal stabilizer and elevator assembly manufactured by Convair/General Dynamics. The entire horizontal stabilizer could be moved in flight for proper trimming. A blue band on the vertical fin was most commonly lettered for MATS, and then MAC; Starlifters in test use sometimes carried test organization abbreviations instead. (Lockheed Martin)

Construction is largely traditional flush-riveted aluminum. Some control surfaces originally used metal-clad honeycomb sandwich material to save weight, but when water seeped into the honeycomb and froze at altitude, it could delaminate this structure. Using more fiberglass in these areas mitigated the problem. The C-141 structure employed the "fail-safe" design rationale, in which a structural failure was not supposed to lead to catastrophic loss of the aircraft.

The C-141's four TF-33 turbofan engines each reside in nacelles hung from forward-projecting pylons beneath the wings. Clamshell thrust reversers, set as left and right halves, can be diverted into the airflow to slow the Starlifter on the runway during landing.

As designed, the Starlifter could replicate sea level air pressure up to 21,000 feet, and an 8,000-foot cabin altitude as high as 40,000 feet.

The Sum of its Parts

There were 1,334 Starlifter subcontractors used in the United States and Canada by the time the first C-141A was rolled out. By weight, 60 percent of the C-141A was produced elsewhere and brought to Marietta for assembly, the Air Force plant representative at Lockheed said upon the first Starlifter's rollout.

The largest of the C-141 subcontracts was with Avco Corporation in Nashville, Tennessee, where the main wing boxes were built and shipped by rail to Marietta, Georgia for mating to the Starlifter. The wings carry all of the Starlifter's nominal 23,000 gallons of jet fuel. Convair/General Dynamics supplied the horizontal stabilizers, spanning 50 feet – greater than the wingspan of small jet aircraft. The Rohr Corporation in Chula Vista, California was responsible for the fanjet nacelles, pylons, and thrust reversers. Rohr's Riverside, California

A newly camouflaged C-141A came in out of the Georgia rain to begin its conversion into a longer C-141B. Lockheed officials had wanted to build longer Starlifters during C-141A production, but the Air Force later decided to stretch existing C-141s and redesignate them C-141Bs, while retaining their original serial numbers. (Air Force)

During the multi-year stretch program, the Military Airlift Command accommodated a mixed fleet of unmodified C-141As and longer, aerial-refueling capable C-141Bs. One of each rested wingtip-to-wingtip at McChord AFB in August 1981, between airlift missions.

C-141A 67-0025 of the 60th Military Airlift Wing was photographed on the ramp at Travis Air Force Base. Gray and white paint scheme characterized C-141As for most of the 1970s. (Travis Air Force Museum collection)

Not really a twin-engine Starlifter, this C-141A had its two outboard engines removed for maintenance when photographed in the early 1980s. Large Fowler flaps are evident in this view.

Before and after photos of the YC-141B stretch modification show the 23-foot fuselage extensions inserted in the B-model. The YC-141B was cut at a different forward fuselage station than the rest of the B-models and used a more extensive wing fillet. (Lockheed)

facility furnished main landing gear doors and pods and petal doors, as well as wing/fuselage panels. Other aero contractors at work on the C-141 included Kaman Aircraft, Moosup, Connecticut, furnishing the tail cone; and Beech Aircraft, Wichita, Kansas, suppliers of Starlifter wing flaps, nose gear door, emergency exits, wing spoilers, and ailerons. Bell Aerosystems of Buffalo, New York, furnished floor plates. Divisions of Bendix Products contracted to provide C-141 main landing gear, the anti-skid system, and the automatic flight control system. The Brunswick Corporation of Marion, Virginia, contributed nose radomes. Cleveland Pneumatic Tool Corporation of Cleveland, Ohio, built the wing flap tracks and the nose landing gear.[14]

C-141A

Emblematic of its status as a simultaneous manufacturing and development program, Starlifter production began with the first C-141A. Traditional notions of starting the series with an XC- or YC-141 were thrown out when the first C-141A, 61-2775, rolled into the hot August sun in Marietta in 1963. Since development was ongoing at that point, the first handful of C-141As included some features absent on later production aircraft. The forward crew entry door hinges outward at the top on the first aircraft, instead of rolling in tracks like subsequent C-141s. The fifth C-141A, identified in the test program as aircraft No. 6005 (The fleet started with 6001, aircraft 61-2775), was to receive all modifications dictated by the flight test program, thereby becoming the exact conformity aircraft for future production. Aircraft 6007 was earmarked for all-weather tests, and 6008 was to be the service suitability aircraft used in airdrop tests and other operational Air Force uses.

The number one C-141A was built with small doors inset in its petal doors that could be opened for airdrop use when the entire aft end of the airplane didn't need to be opened up in flight. Although this was not a production feature, some C-141As entered airlift service with vestiges of these openings skinned over.[15]

Through much of the 1960s as C-141As were built in Georgia, upgrades treated problems with landing gear cracks and pressure bulkhead failures as the Starlifter matured. Yet overall, appearance of the first C-141A is very similar to the last aircraft that left the Georgia assembly line on February 27, 1968. All Air Force Starlifters were initially C-141As; stretched B-models were all conversions, and upgraded C-models were improved-upon C-141Bs already in the fleet. The early C-141As with outward-opening crew hatches did not serve MAC as cargo haulers, but became testbeds within Air Force Systems Command, ending their service in the late 1990s with the follow-on Air Force Materiel Command at Edwards Air Force Base, California. To indicate their nonstandard use, the testbeds were additionally classified as NC-141As, with the "N" denoting these Starlifters were not suitable for their original mission.

When production of 284 Air Force Starlifters and one civil L-300A variant ended, the run was possibly foreshortened by as much as a third because of the impending arrival of the C-5A. As early as 1964, the Air Force suggested to the Department of Defense that procurement of eight squadrons of C-5s, at that time still known by the acronym CX-HLS (cargo, experimental – heavy logistics system) could result in a smaller C-141 buy than the approximately 390 Starlifters that had been suggested. Had Starlifter construction extended to 390 aircraft, at a nominal production rate of seven C-141s a month, delivery of C-141s could have overlapped delivery of C-5As from the same Lockheed plant.[16]

The C-141B stretch program yielded more cargo capacity for MAC with no additional airframes bought.

When the demands of airlift necessitated returning a newly stretched C-141B (65-0248) to service before it could be painted, the result graphically depicted the surgery performed to make longer Starlifters. Paint silhouette in the shape of the landing gear pod shows how far aft the fuselage was moved.

In July 1981, C-141B 65-0248 returned to service before it could be scheduled for painting following completion of stretch modifications. Two small discs on the side of the fuselage show where the leading edge wing scanning light formerly was (forward circular patch) and its replacement location in the new stretched section, to retain its original angle to the wing.

MAC recorded its highest number of Starlifters in service, 277 C-141As, in 1968 and 1969. Attrition, plus the test aircraft that did not enter the airlift stream, account for the rest of the production tally.[17]

C-141B

The two features most often associated with the C-141B – longer fuselage and aerial refueling capability – had been considered and rejected by the Air Force in May 1961, more than two years before the first aircraft was built. MAC historical documentation indicates the decision to delete a requirement for aerial refueling on the C-141A was made on May 18, 1961, along with the deletion of a proposed forward cargo door on the side of the fuselage. A week later, according to the MAC history, "The possibility of lengthening the cargo compartment for increased capacity was unfavorably considered since such action would necessitate complete redesign and extensive additional wind tunnel testing with attendant increased costs."[18]

Lockheed had long championed a stretched version of the Starlifter. In the mid 1960s, the company suggested that the original length of the C-141A could be increased by 37 feet, yielding a much longer Starlifter riding on triple bogie main landing gear.[19] The actual stretched C-141B was 23.3 feet longer than the A-model from which it came, and retained conventional Starlifter landing gear.

When Israel became embroiled in war with Syria and Egypt in October and November 1973, European allies refused landing rights for American aircraft intended to resupply Israel. Though ultimate access to the Azores enabled MAC C-141As to reach the Middle East, the sobering reality of airfield denial blunted the ability of the U.S. to use C-141 airlift as a rapid-response foreign policy tool. Air Force historian William Head suggests this experience gave new momentum to the cause of providing Starlifters with aerial refueling capability.[20]

When funds were allocated to Lockheed by the Air Force in 1975 to convert one C-141A to the longer, aerial-refuelable, B-model configuration as a test, the service initially weighed the option of having production stretch modifications made by the Air Force's Warner-Robins Air Logistics Center, or other Air Force logistics sites, or by a bidder to be chosen in competition. Lockheed ultimately won the nod to make production Starlifter stretches as sole-source contractor; logical arguments were put forth for having the Starlifter's original manufacturer be the company to stretch its creation. Among the compelling reasons for having Lockheed stretch Starlifters

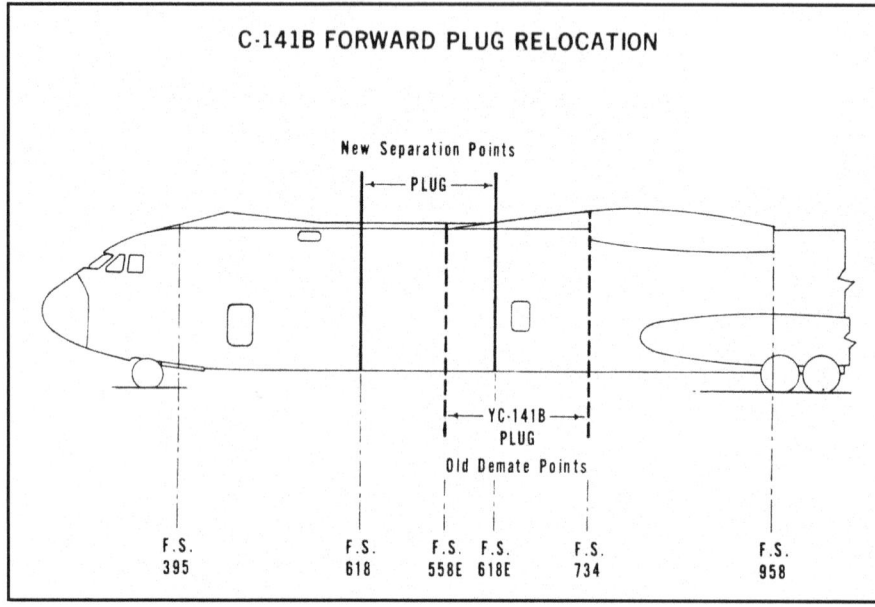

The location for splicing new structure into the forward fuselage to make B-model C-141s was moved after Lockheed stretched the YC-141B prototype. (Warner Robins Air Logistics Center History Office monograph)

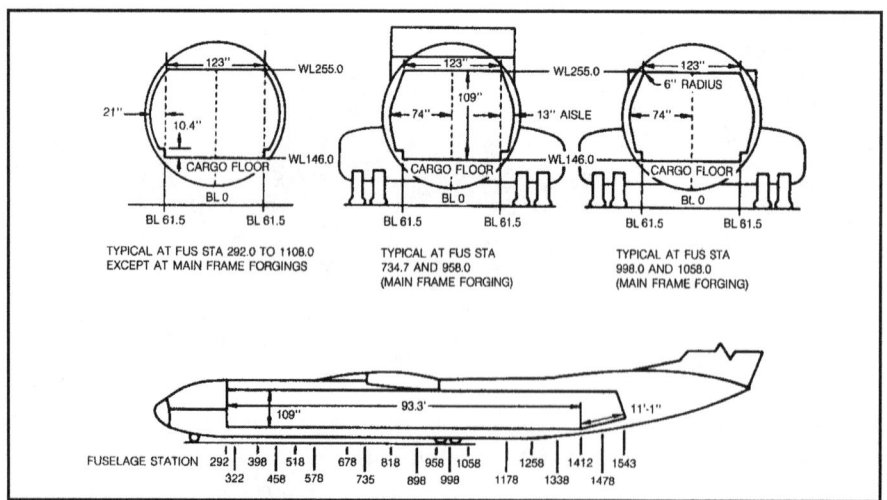

C-141B cargo compartment drawing illustrates clear loading space obtained by mounting the wing high on the fuselage. (Lockheed Martin)

A 452nd Air Mobility Wing C-141C from March Air Reserve Base, Riverside, California flew into the 21st century as Starlifters played out a collective career spanning more than four decades. This C-model was photographed departing Christchurch, New Zealand. (Air Force photo by TSgt. Rick Sforza)

Massive C-141 wing box assemblies were manufactured by Avco in Tennessee and shipped by rail to Georgia for final assembly. (Air Mobility Command Museum collection via Harry Heist)

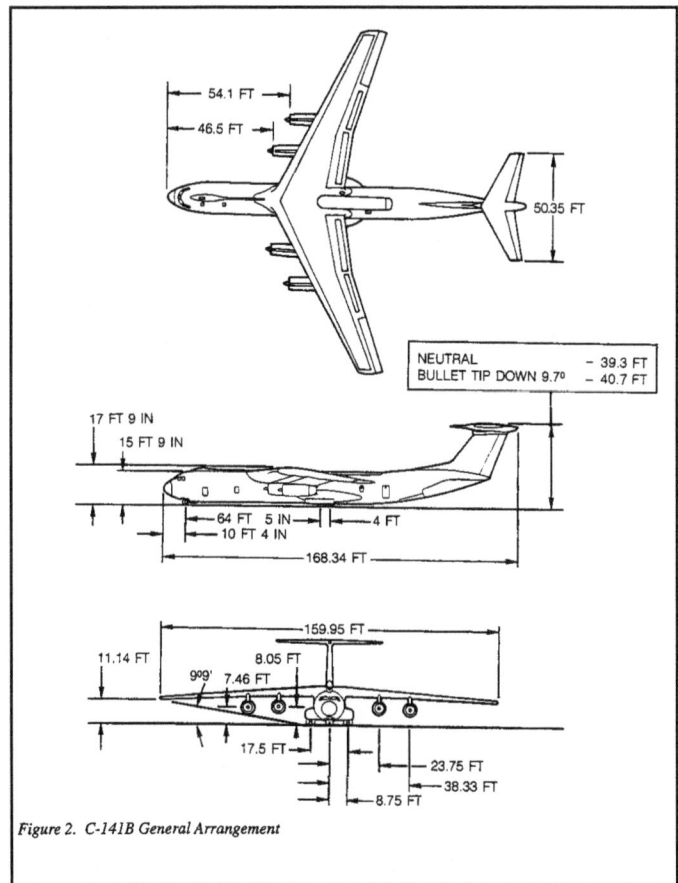

C-141B general arrangement three-view drawing shows variable measurement for Starlifter height depending on position of movable horizontal stabilizer. (Lockheed Martin)

was the company's previous research into the effects of loads stress on the C-141's airframe.

The Air Force's Warner Robins Air Logistics Center, near Macon, Georgia, would paint the stretched C-141s and perform periodic maintenance on them. The decision to have the C-141B modifications performed by Lockheed was made in the shadow of controversy stemming from cost overruns of another Lockheed Air Force transport, the giant C-5A Galaxy. The Air Force convened a Should Cost Study Team to scrutinize Lockheed's Starlifter stretch proposal.[21]

One point of contention between the Air Force and Lockheed was the plan by Lockheed to install new, more extensive wing fillets during the stretch modification. The Air Force questioned the benefits of the new wing fairings when balanced against the cost. Flight test reviews in August 1977 of the YC-141B fitted with the new fillets suggested the benefits were not as great as anticipated, according to a Warner Robins Air Logistics Center history of the program. Lockheed designed the fillets to aerodynamically reduce wing movement, and hence extend wing life. Under test and evaluation, the Air Force decided that while the new enlarged fairings did promote fuel efficiency, the stated goal of achieving a 45,000-flying-hour life for Starlifter wings was attainable with the original small A-model wing fillets, too. The cost of new fillets, plus their added weight, at 1,500 pounds, argued against their incorporation.[22]

Ultimately the Air Force chose to retain smaller C-141A wing fillets. When Lockheed created the concept for stretching the C-141, including the use of big fillets, the forward fuselage

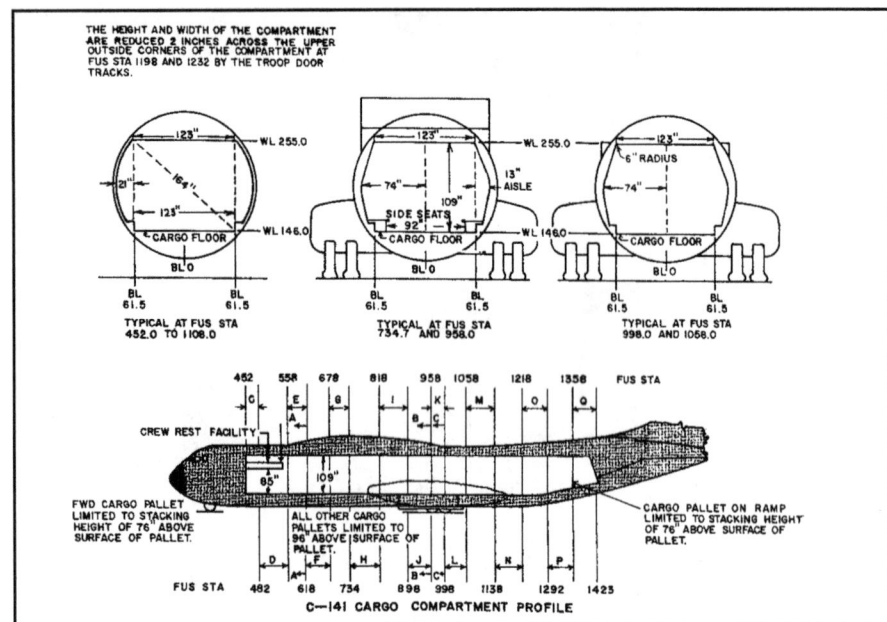

Drawing of C-141A cargo compartment shows ability to carry a pallet of cargo on the loading ramp, albeit with height restricted to 76 inches above the pallet surface to accommodate the tilt of the ramp in the closed position.

splice was intended to be at fuselage station 734. Though feasible, this location required extensive rework of electrical and hydraulic systems at that location. Along with the demise of the enlarged wing fillets, the Air Force requested the forward fuselage splice be relocated to fuselage station 618 to simplify the labor involved. As a result, the YC-141B was unique among stretched Starlifters; only it had the enlarged wing fillets and the forward fuselage splice at station 734.[23]

The stretch modification added about 13 feet ahead of the wing and 10 feet behind, since the fuselage had to be stretched in front of and behind the wing to retain an adequate overall center of gravity. The forward splice on production C-141Bs at fuselage station 618 was

Mass airpower congregated at Dover Air Force Base, Delaware in March 1982 for the Gallant Eagle airdrop exercise. Army troops and equipment were flown cross-country nonstop to Fort Irwin in California's Mojave Desert, where they were airdropped. Starlifter TF33 engine nacelles, built by Rohr under contract, use heat to keep the cowl lip from icing up. Over time, the heat could cause gray painted inlets (usually on European One camouflaged aircraft) to shed some paint in patches. Later, cowl inlet rings were left unpainted.

Early Lockheed C-141A flight tests included instrumentation on tips of wings and tail. (Lockheed Martin)

complemented by an aft splice at station 1058, avoiding an area of compound curves to the rear. As with the forward splice, the aft cut was made in an area of the fuselage where as few systems as possible would be disturbed by the severing. Slight modifications were required to the aft fuselage behind the separation point to match the new structure, which placed the aft fuselage well behind the wing. Differences in aft plugs between production B-models and the YC-141B resulted when the enlarged wing fillet was discarded.

Relocating the forward splice closer to the nose of the C-141 allowed lighter construction to still safely accommodate loads on the airframe. This had the additional benefit of saving 500 pounds in weight per airplane. As built, production C-141Bs weighed about 8,000 pounds more than a short A-model; the heavier YC-141B with enlarged fillets and different forward splice location weighed about 10,000 pounds more than a C-141A.[24]

The stretched version had an additional 233 square feet of floor area, resulting in more than 2,100 cubic feet of additional useful space in the cargo compartment. Three additional cargo pallets could be loaded in this space. Since C-141As sometimes filled with bulk cargo that did not reach the aircraft's maximum weight, the longer

The first of the breed, C-141A, construction number 6001 (Air Force serial 61-2775), as it appeared circa January 1963 during fuselage section mating. The main landing gear pod fairing is visible in the right of the photo. (Lockheed Martin)

models could carry more bulk; in such conditions, three C-141Bs can do the work of four C-141As. As the Air Force described it in a fact sheet: "Modification of the A-models produced the equivalent cargo capacity of 90 additional A-models, all at a price well below that of purchasing new aircraft. Also, no additional crews or support equipment are required, further cutting costs for the increased cargo space." Not mentioned was the inescapable math: the crash of one stretched C-141B would remove more than one C-141A-equivalent from airlift availability.

In addition to the obvious fuselage plugs and aerial refueling plumbing atop the fuselage, the B-models incorporated titanium straps to strengthen the center fuselage because of the increased bending loads.

The first stretched Starlifter, designated YC-141B, flew to the Air Force Flight Test Center at Edwards Air Force Base, California on June 6, 1977. Flying qualities, airdrop, landing, minimum control speed tests, and air-refueling capability tests were conducted there. The air-refuelable B-model is said to be able to transfer jet fuel at about 900 gallons a minute.

Following success with the prototype stretched YC-141B, in June 1978 Lockheed-Georgia garnered an Air Force contract worth $407.5 million to stretch the MAC Starlifter fleet, then listed as 271 C-141A-models. The total number of aircraft finally receiving B-model modifications was 270. By October of 1980, the C-141 modification line attained its peak efficiency. At that time, a C-141A emerged as a C-141B after 64 workdays in Georgia; with more than one Starlifter undergoing conversion at once, the flow was ongoing.[25]

With alignment fixtures keeping sections in close tolerance, the new

The first flight crew of the C-141 on December 17, 1963 was photographed in orange flight suits with Air Force representatives. Crewmembers, from left to right, were: A.P. Bob Brennan, flight engineer; E. Mittendorf, flight test engineer; Henry Dees, co-pilot; and Leo Sullivan, Lockheed-Georgia chief engineering test pilot. (Lockheed Martin)

Three newly coined C-141As receive final attention outside the Lockheed plant at Marietta, Georgia. Hard engine inlet covers seen in early photos gave way to fabric covers that could be stashed in the landing gear pods. The Starlifter nearest the camera is construction number 6084 (Air Force serial 65-0233).

Eleven Starlifters undergoing stretch modifications are visible in this assembly line photo, giving an indication of the massive size of the Lockheed-Georgia plant. (Lockheed Martin)

fuselage plugs were mated to the fore and aft fuselage sections of cut-apart C-141As before these elongated pieces were joined to the fuselage center section. During production stretch operations, it was apparent that the incoming C-141A aircraft could be significantly different from each other because of previous repair histories. A detachment from the Air Force's C-141 Warner Robins depot moved to Marietta to keep Lockheed stretch program workers appraised of the quirks of each individual C-141A.

Lockheed earned a reputation for delivering B-models ahead of schedule. The expanding Lockheed production line was so efficient that for a while in 1981, the number of C-141s available for MAC airlift on a given day dropped from 152 to only 127. This was exacer-

A dozen Starlifters share space with new C-130s at Lockheed-Georgia during conversion to C-141B configuration. (Lockheed Martin)

bated by consecutive depot maintenance time for C-141s at Warner Robins. Using other airlift aircraft, plus shortened maintenance periods at Warner Robins, helped alleviate the Starlifter shortfall effects. This may explain unusual events like the temporary return of a stretched 62MAW C-141B to MAC service with the fuselage extensions unpainted.

The trouble-free success of the B-model conversion program prompted the Air Force and Lockheed to begin other upgrades to Starlifters concurrent with the stretch modifications. These included new weather radar and station keeping equipment (SKE) that facilitated formation airdrops.

When the 270th and final C-141B was delivered to the Air Force on June 29, 1982, Gen. James P. Allen, MAC commander, alluded to the nickname of the legendary and timeless C-47 transport when he said: "The era of the C-141A ends today, but these 'gooney birds' of the 1980s usher in a new era in air mobility." The C-141B program finished two weeks early, and $20 million under budget.[26]

Flaps promoted a shallow landing approach for this early C-141A at Lockheed-Georgia. (Lockheed Martin)

This C-141B undergoing wing removal shows the general contour of the wing root airfoil section. (Lockheed Martin)

Elements of the aerial refueling receptacle are parked to the left of the C-141 undergoing stretch modification at Lockheed-Georgia. Production stretch mods involved a splice at a different forward station than the first YC-141B. (Air Mobility Command Museum collection)

The number one C-141A (61-2775) photographed with its radome hinged open for servicing at Wright-Patterson Air Force Base, Ohio in March 1984. The Starlifter's cockpit side windows can be opened.

During maintenance in the spring of 1984, the first Starlifter (61-2775) was photographed with hatches and panels opened or removed for inspection and servicing. Parallel black lines down the spine of the fuselage outline the service walkway.

The Numbers

A mid-1979 table produced in MAC compared a number of measurable features of C-141As and longer C-141Bs:

	C-141A	C-141B
Maximum Ramp Weight	325,000 lbs	325,000 lbs
Maximum Takeoff Weight	323,000 lbs	323,000 lbs
Air Evacuation Capacity (Ambulatory/Litters)	131/79	131/103
Passenger seats available (Overland/Overwater)	152/133	200/153

(Wartime contingency operations could exceed these weights for both models by 10 tons. Some numbers may have been rounded off in this table; aircrew members may recall a maximum takeoff weight specifically 100 pounds higher than that listed in this general chart.)

C-141C

In the 1990s, 63 C-141Bs were earmarked for upgrading with glass cockpit instrumentation suites. The C-141 autopilot and cockpit upgrade (ACUG) program was followed by a test of the global positioning system enhanced navigation system (GPSENS). ACUG tested a new all-weather flight control system (AWFCS); aircraft subsequently modified with CRT screen instruments (glass cockpit), new autopilot, and GPSENS were differentiated by the nomenclature C-141C. The process of converting C-141Bs to C-141Cs began in 1997 and finished in 2001. Deployment of C-models started in October 1997.

The C-models were delivered to Air Force Reserve and Air National Guard units as active-duty airlift wings increasingly filled their ranks with new C-17s. By the spring of 2004, a

When photographed in March 1984, the first Starlifter still retained small paradrop doors inset into the larger petal doors. Though deleted as a general-production option, vestiges of these small doors remained for years in the form of some petal doors delivered with skin patches where the inset doors were to have been.

A side view of testbed NC-141A 61-2777 shows upward flex of wings with flight loads. Cylindrical aft fuselage extension accommodated various avionics packages for test purposes. This photo was taken in September 1994 on the last flight of 2777, from Edwards Air Force Base to the storage facility at Davis-Monthan Air Force Base, Arizona. (AFFTC History Office)

First flight of the stretched YC-141B (66-0186) took place in March 1977 over the red clay hills of north Georgia. Visible in the shadow of the wing is the larger fillet used on the YC-141B, but not adopted on the rest of the B models. (AFFTC History Office)

reduced fleet of C-141Cs, intended to serve until their announced retirement in 2006, boasted the glass cockpit with AWFCS and a ground-collision avoidance system (GCAS).[27] By June 2004, Lockheed Martin counted 24 C-models still in service with 39 retired, including one listed as attrition after a ground mishap.

Abundant Power

If the World War II P-51 Mustang famously benefited from its employment of the Merlin V-12 engine, the C-141 Starlifter owes a measure of its luster just as surely to the development of the Pratt and Whitney TF33-P7 fanjet engine, rated at 20,250 pounds TRT (takeoff rated thrust). MAC historians considered the mar-

Clad in various protective coatings, a Starlifter under construction was shown to the public during Lockheed-Georgia's annual open house on Armed Forces Day, May 18, 1963. (Photo by Kenneth G. Johnsen)

riage of the C-141 airframe to the TF33 engine key to the aircraft's success: "...The TF33-P-7 clearly had no equal at the time and made possible a unique transport aircraft in the C-141. By the time the Military Air Transport Service had acquired a new name on 1 January 1966 as the Military Airlift Command, it had also acquired a new image as a vigorous and progressive command."[28]

Col. William H. Spillers, C-141 program manager from 1959 to 1963, told MAC historians the Starlifter marked "the first time we really had an airplane that had all the thrust and power that it required. In fact, it probably was a little bit excessive if you compare it with older airplanes..."[29] In 1935, when Boeing unveiled its four-engine Model 299 Flying Fortress prototype, that new bomber's remarkable performance was enhanced by engines of sufficient power and quantity to set speed records. With the C-141, again an airframe was mated to engines of sufficient power and quantity to ensure favorable performance instead of falling into the designer's trap of seeing how large an airframe could be hefted by the available powerplants.

Not only did the TF33 possess great power, it also enjoyed durability in service. As the engine amassed flight hours, it was tested to determine how long it could fly between overhauls, a figure that grew as the TF33 proved itself. The figure for time between overhaul (TBO) for Starlifter engines was initially set at 2,400 hours in 1964. As engines were inspected and analyzed, the TBO was nudged up to 3,200 hours, and incrementally

Lockheed-Georgia workers looked on as a C-141A tail assembly from Convair took shape on the Starlifter assembly line. (via Harry Heist, Air Mobility Command Museum collection)

Top view of the original YC-141B shows larger wing fillets than used on production Starlifters. White refueling housing above the cockpit would later be painted gray. (AFFTC History Office)

The seventh C-141A built (63-8076) came in for a landing at Dobbins Air Force Base, Georgia, on July 5, 1964, with grass and pavement reflected in its shiny natural metal skin. (Photo by Kenneth G. Johnsen)

higher thereafter. In 1967, six C-141 Lead-the-Force accelerated-use aircraft were designated to operate 4,000 hours between engine overhauls. By early 1970, the C-141 TF33 TBO was 8,000 hours, significantly boosting its utility and economy. Refitting engines with improved replacement parts including first-stage turbine wheels, transition ducts, and carbon seals promised to deliver a TBO of 12,000 hours for engines so modified. Carbon seal degradation was alleviated by using engine oil better suited for TF33s. Engine monitoring was instituted in MAC in an effort to predict engine failures and to stretch the TBO of engines that did not show performance anomalies based on frequent logging of performance statistics in a database.

To support the Starlifter fleet, the Air Force bought 1,468 TF33-P7 engines from Pratt and Whitney for a total of $390,981,284, yielding a unit cost of $266,336. The supply of engines, including four per C-141 spread over 284 Starlifters, also yielded 332 as spares. As Starlifter flying hours grew, a discrepancy arose between engines on operational airlifters and those flown on training Starlifters. Airlift C-141s typically flew longer en route sorties at cruise conditions than did their training counterparts. This resulted in higher "cycles" for engines on training C-141s. Cycles describe the sequence of events from engine start-up through takeoff, cruise, landing, and shutdown, when variations in temperature and RPMs can adversely affect engine components. The Starlifter fleet was accumulating airlift engines with fewer cycles and more overall flight hours, and training engines with high cycles but lower flight times. To even out the wear-and-tear on the TF33 fleet, a deliberate process of swapping engines between air-

The Starlifter nosewheel door drops down and aft to allow wheels to extend and retract into the fuselage.

Details of the outward-blowing APU exhaust, and gear door hinges, are visible in this July 2004 photo of the left landing gear pod of the C-141C Hanoi Taxi, repainted gray and white in honor of its historic service in those colors as a carrier of POWs in 1973.

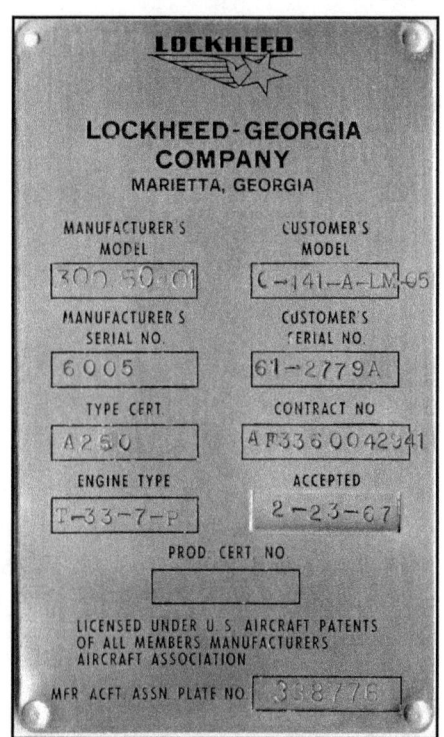

With minor digital restoration to this image for clarity, this is the Lockheed data plate from NC-141A 61-2779, in the collection of the Flight Test Historical Museum at Edwards AFB, California. (Courtesy Doug Nelson, Flight Test Historical Museum)

lifters and trainers was instituted in September 1969.

The basic TF33 turbofan (civilian designation JT3D) was derived from the older turbojet J57. To achieve the JT3D/TF33 turbofan configuration, the first three J57 compressor stages were deleted and two fan stages were installed. The fan portion of the TF33 is substantially larger in diameter than the rest of the engine. Other changes to the basic J57 to create the TF33 included enlarging the third-stage turbine and adding a fourth stage to power the low-pressure compressor rotor and integral fan. The J57 ancestry of the TF33 was further underscored when some J57s (JT3C civil versions) were converted into JT3D turbofans. Compared to J57 turbojets delivering maximum takeoff power in the range of 13,500 pounds of thrust, the TF33-7's 21,000-pound rating put the Starlifter well ahead of its J57-powered airline contemporaries. Flight trials of this promising turbofan began in 1960 using a B-52 and a Boeing 707 jetliner as testbeds. The JT3D was subsequently adopted in versions of the 707 as well as the competing Douglas DC-8, and the B-52H. In 1973, a further refined variant, the TF33-PW-100A, was picked to power Boeing's E-3A AWACS derivative of the 707.

Production of the JT3D/TF33 family of turbofan engines exceeded 8,550 copies delivered through 1984.[30]

WHAT MIGHT HAVE BEEN AND LOOK-ALIKES

Reflecting its C-141A heritage in its U.S. civil registration number – N4141A – the only Starlifter built for civilian purposes was campaigned by the company in an effort to make good on one of the aircraft's original premises: joint civil/military usefulness. Ultimately, short A-model Starlifters did not generate civil orders, and even a proposed stretched version was not ordered in sufficient quantity to make production economical. (Lockheed via Craig Kaston collection)

rated to lower power settings to provide lower operating costs and a longer time between overhauls.[31]

Efforts to make this a multi-purpose military and civil transport predate the first flight by several years. As early as December 1959, the Federal Aviation Agency (FAA) polled air carriers about desirable cargo aircraft specifications. Communications from 25 airlines, offering a total of 29 sets of specifications, allowed the FAA to devise a cross-section of civil cargo aircraft requirements, which the agency gave to the creators of Specific Operational Requirement (SOR) 182. SOR-182 informed the design of the Starlifter as a military transport, and gave rise to the competition among several prime contractors to win the Air Force contract to build what became the C-141.

SOR-182's provisions included the requirement that "...the aircraft

The Solo Civil Starlifter

One Starlifter was campaigned by the company as a demonstrator of commercial cargo utility as Lockheed Model 300A. In an ironic twist on traditional transport designs, the Model 300/C-141 initially met the requirements of the Air Force, while simultaneously attempting to address concerns of potential airline operators. The Starlifter was not a rehashed civil transport.

While past transports may have labored under powerplants that were merely adequate, some civil carriers suggested the Starlifter's remarkably powerful TF33 engines were too costly for effective commercial operations. Lockheed countered by suggesting that the TF33s could be de-

Lockheed's sole civil Starlifter demonstrator carried varied cargo, including an engine for the C-5A, as the company worked to interest commercial carriers in the L-300. (Lockheed Martin)

Given NASA registration number N714NA, the civil Starlifter embarked on a second career as an airborne telescope platform in the first half of the 1970s. (NASA photo via Craig Kaston)

NASA's white and blue Starlifter served as an airborne observatory for more than two decades. Like the handful of testbed NC-141As, this Starlifter did not receive the stretch modification.

Substantial structural modification was necessary to install an opening door in the fuselage of the NASA Starlifter. This new door allowed the telescope to be used while the rest of the pressurized cabin and cargo compartment was preserved.

shall be designed and shown to comply with Civil Air Regulations, transport category, at the civil payload, range and field size specified." The FAA administrator from 1961-65, Najeeb Halaby, is quoted in a letter as saying, "...civil certification is a condition of the Air Force contract with Lockheed." Air Force Regulation (AFR) 80-36 spelled it out: "Transport aircraft that the Air Force procures or develops must be designed to comply with civil airworthiness standards when their intended usage is generally consistent with civil operations." The pertinent AFR did not preclude "the use of military specifications and standards in designing an aircraft when necessary to assure that the aircraft will perform its military role properly under intended operating conditions." The regulation also instructed, in the case of a design conflict between military and civil mission requirements, "...first emphasis will be placed on meeting the military mission requirements." A March 1961 memorandum between the Air Force and the FAA established a permanent FAA presence within the C-141 program.[32]

The dual civil and military roles forecast for the Starlifter design led to at least one interesting extreme. The Air Force only required the landing gear to be designed to tolerate sink (ground contact) rates of nine feet per second (fps). The FAA required 10 fps strength in the design, so the more stringent FAA requirement was met, which automatically accommodated the Air Force's need. The FAA wanted braking limits to be achievable without using thrust reversers, since the time lag in using thrust reversers could make them negligible in affecting required landing distance calculations.[33]

Some in the airline industry argued that the Department of Defense (DoD) and the FAA did not accord airline requirements enough

Lockheed had the paint and markings ready to go in concept form when cargo carrier Flying Tiger Line put money down for civil stretched L-300B Starlifters. Unfortunately, lack of orders doomed the elongated civil L-300B. (Lockheed via Flight Test Historical Museum collection)

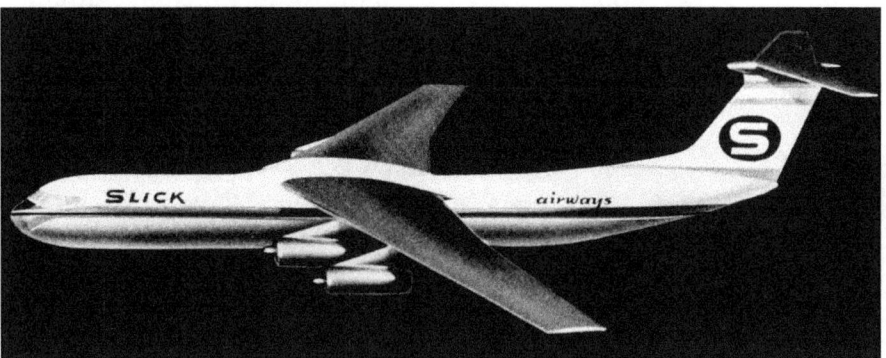

Slick Airways put money down on Lockheed's proposed lengthened L-300B civil Starlifter, but the orders for Slick and Flying Tiger went unbuilt. Lockheed had hoped the Air Force would order elongated C-141s from the assembly line to make the unit cost worthwhile, but it was several more years after the end of C-141 production before the Air Force contracted to stretch its existing short C-141As. (Lockheed via Flight Test Historical Museum collection)

consideration when drafting SOR-182 in 1960, but Lockheed remained optimistic that civil Starlifters could operate economically. At the start of C-141A flight testing, Lockheed was reported to have estimated that its cost for dual military and commercial certification of the Starlifter would run to $30 million.

Early preflight airline skepticism included concerns that the C-141A's ability to operate from 6,000-foot runways would necessitate weight or design compromises that could boost commercial operating costs. By 1961, Lockheed told potential airline customers the L-300A commercial variant would weigh less than a military C-141 by 2,800 to 3,600 pounds. The weight savings came from deletion of military equipment including the hydraulics involved with air delivery, military loading equipment, and commercial suitability of a lighter floor than was required by the Air Force.[34]

Though pitched primarily as a civil cargo aircraft, the Lockheed L-300A was groomed to receive FAA Type 4B certification, enabling the carriage of passengers as well as cargo. FAA engineering test pilots were involved in C-141A flight testing to expedite the certification process. On January 29, 1965, the Starlifter received its type certificate from the FAA in ceremonies at Dulles International Airport outside Washington, D.C.

Early Air Force forecasts of a 132-airplane C-141 purchase were lower than the ultimate buy, but based on these figures, Lockheed officials expressed hope that a significant commercial fleet of L-300s could be made available to augment military airlift in times of national emergency.[35] Years later, airlines participating in the Civil Reserve Air Fleet – CRAF – supported the massive airlift operations of Desert Shield and Desert Storm in 1990-91 with transports including 747s and DC-10s.

By 1964, Lockheed was campaigning a promised elongated version of the commercial L-300A Starlifter, the L-300B Super Starlifter. Lockheed touted the stretched L-300B proposal in 1964 promotional literature: "The L-300B is seen as the key factor in an anticipated seven-fold growth in air cargo during the next decade. The airplane's operating efficiencies and performance will cut ton-mile costs drastically, while allowing the industry to compete with surface transportation and carry a wide variety of freight now traveling by other means." Specifications for the proposed L-300B included a cargo hold "104 feet in length, 25% longer than that of the C-141." This is at odds with other L-300B figures suggesting a 23-foot stretch over the C-141A's 70-foot compartment, possibly indicating differences in configuration or use of the ramp area in computing cargo area. The specifications also said: "The L-300B will operate from medium size fields at maximum take-off weight, 366,000 pounds."[36]

Lockheed also referred to a proposed civil Starlifter 23 feet longer than a C-141A as the "Lockheed 301" in a February 1965 in-house article.[37] Studies suggested the lengthened L-300B could achieve profitability at a freight rate of 10 cents a ton mile, which was considered to be favorable to then-current surface truck rates of

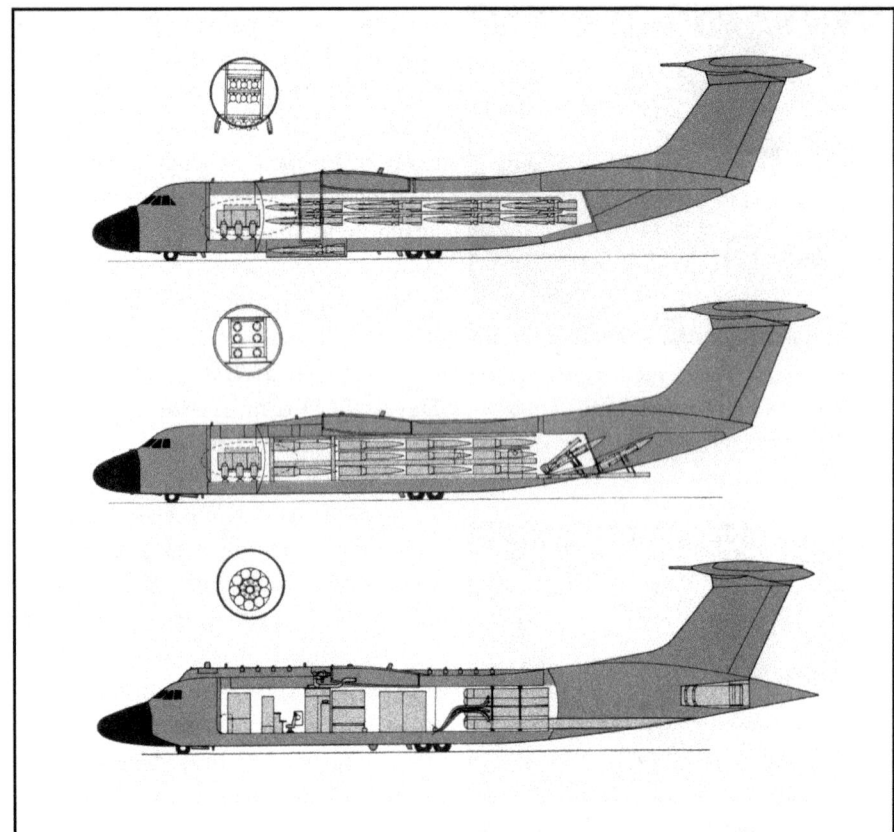

It's a tribute to the versatility of the C-141's airframe and the flexibility of Lockheed designers that the Starlifter was proposed as an air defense missile launcher in several big-nosed configurations. One used a forward-launching missile tray; another was envisioned to carry 22 missiles to be launched out the back. An antenna-studded variant with elongated tail cone featured a revolving missile cylinder. The missiles were to be launched aft pneumatically before firing to seek their targets. Though it wasn't built, it wasn't for lack of trying. (Lockheed-Georgia via Craig Kaston collection)

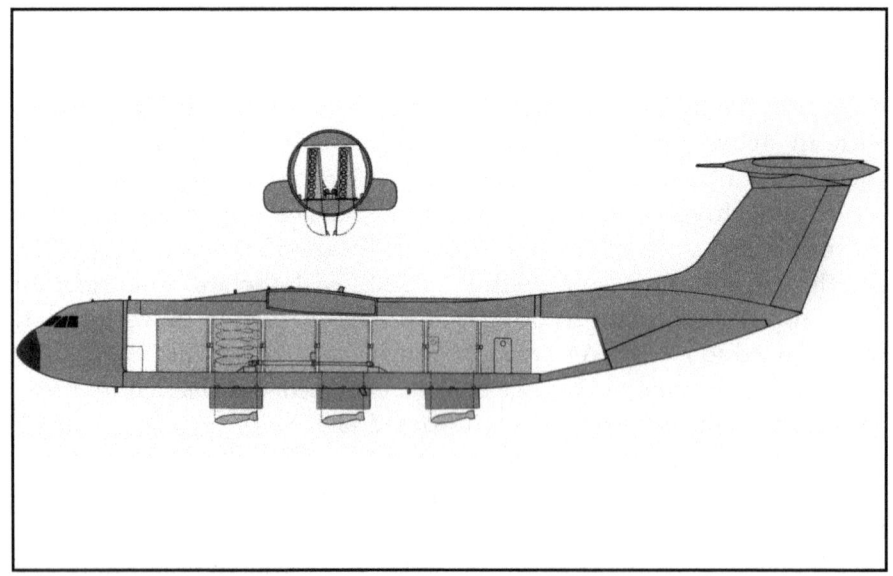

Company promotional artwork depicted a potential Starlifter bomber in the last half of the 1960s, releasing gravity bombs from three bays in tandem. (Lockheed-Georgia via Craig Kaston collection)

eight cents a ton mile. Long before any stretched Starlifters emerged from the plant, air cargo hauler Slick Airways reserved four delivery slots for L-300Bs, followed in May 1964 by a $500,000 deposit from the freight-carrying Flying Tiger Line on 12 stretched Starlifters, worth $64 million. One attraction to purchasing commercial Starlifters by all-freight airlines was the ready compatibility with Air Force C-141s that could enhance the civil carriers' chances of winning contract flights from MATS.

Both orders included stipulations that enough stretched Starlifters would have to be ordered by users to justify production, a figure some industry observers put at 75-100 aircraft. Lockheed wanted the Air Force, then the purchaser of 132 shorter C-141As, to make any additional C-141 purchases include a quantity of stretched variants. In the end, the Air Force ordered only new C-141A models; the stretched Air Force C-141B model was a later rebuild of existing airframes. A dearth of orders for civil Starlifters of any length doomed the sale of commercial L-300s, blunting the promise of one of the Starlifter's touted developmental premises.

By 1964, Lockheed was using its civil-marked L-300A Starlifter in advertising campaigns intended to lure customers, saying: "Cross-section of the Lockheed 300's rear opening measures 9 feet high by 10 feet wide. This means straight-forward loading of standard transportation units, such as 8x8x10-foot containers or pallet loads with straight sides and flat tops up to 9 feet high preloaded at shipper, forwarder, or airline terminals." By emphasizing the size and volume of the containers a Starlifter could haul, and how they could be pre-filled for quick insertion into the aircraft, the advertisement delineated the Starlifter's advantages over contemporary airliners used for cargo. And yet, the short-bodied L-300A did not take business away from other jetliners used for freight.

The Lockheed civil Starlifter demonstrator carried cargo on occasion. In 1966, Lockheed-Georgia and Lufthansa sponsored the airlift of two Bell UH-1D helicopters from the Bell plant in Fort Worth, Texas to Munich, Germany. The nonstop flight lasted nearly 10-1/2 hours. In the summer of 1967, when Middle East turmoil temporarily closed the Suez Canal, Trans-Mediterranean Airways of Beirut leased the Starlifter, wearing U.S. civil registration N4141A, to airlift a pump station for crude oil from Amsterdam to Iran.

The sole civil Starlifter took on a second career as a NASA asset to mount a unique airborne telescope. In 1972-73, Lockheed facilities at Ontario, California, modified the Starlifter, enabling its use as the world's largest airborne astronomical observatory. Central to the success of this venture was the Starlifter's ability to fly anywhere in the world to place the telescope in the best location for astronomical observations. Able to operate from 37,000 to 45,000 feet above the earth, the Starlifter placed its telescope above 99 percent of the atmosphere's interfering water vapor, and above 85 percent of the total atmosphere, surmounting problems that hindered terrestrial observatories.

The core of the Starlifter observatory was a 36-inch telescope weighing seven and a half tons. A special pressurized structural partition in the left side of the fuselage ahead of the wing root enabled the telescope's temperature to be controlled to match the outside air, reducing shocks to the system when a one-of-a-kind door in the top of the fuselage opened to allow the telescope access to the heavens. To accommodate the opening in the pressurized Starlifter, Lock-

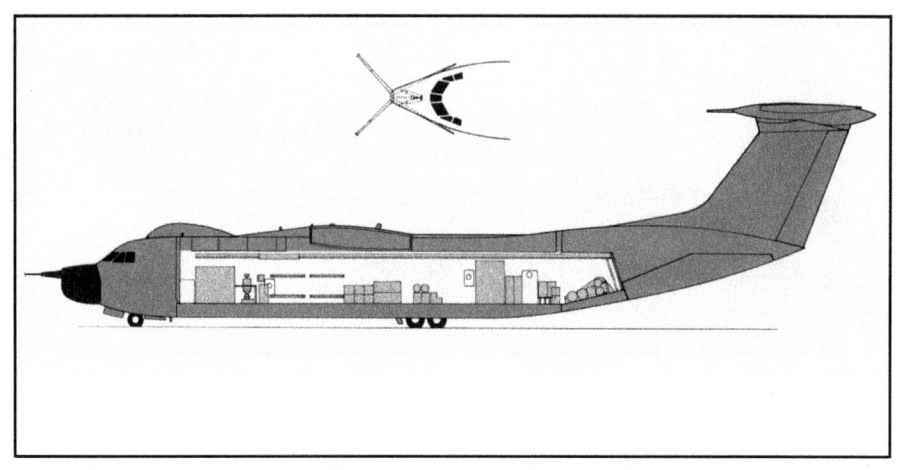

Capitalizing on the success of HC-130H Hercules aircraft, which could extract a downed flier needing rescue from a remote location by snaring a balloon-borne line and winching the person aboard, Lockheed suggested a Starlifter variant could do the job and bring the rescuee home more quickly, too. Though not built, the Starlifter concept closely followed the HC-130 rationale. (Lockheed-Georgia via Craig Kaston collection)

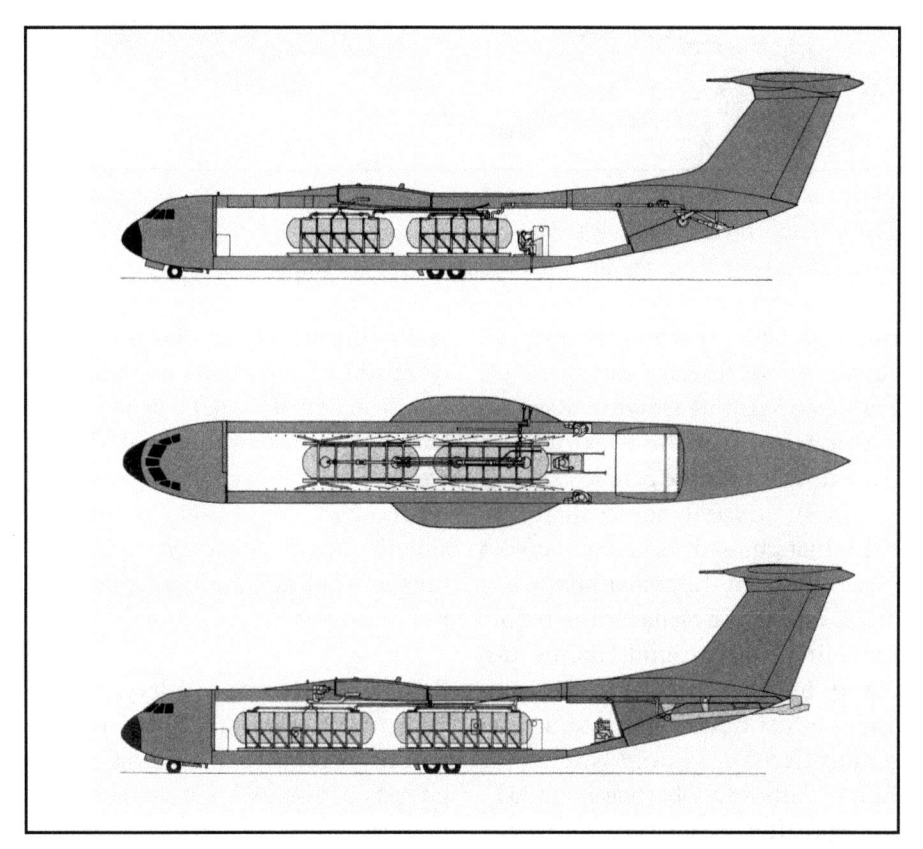

With a station for a boom operator inside the pressurized cabin, three proposed Starlifter tankers would have used a variety of boom and probe-and-drogue equipment to service other aircraft in flight, as depicted in promotional materials. A minimalist version (top) could be reconfigured for transport missions, Lockheed said. A dedicated probe-and-drogue version (middle), with a central boomer and two side scanners, was compared to the slower KC-130F tanker version of the Hercules. The flying boom variant (bottom) was said to be designed to tank 66,920 pounds of fuel for a 2,370-nautical-mile mission radius. All three remained on the drawing board. (Lockheed-Georgia via Craig Kaston collection)

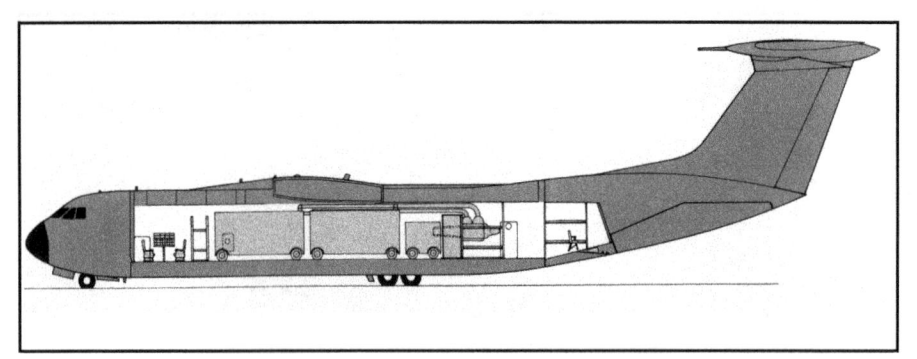

Mobile communications vans were envisioned inside a Starlifter communications platform aircraft in company promotional artwork from the 1960s. (Lockheed-Georgia via Craig Kaston collection)

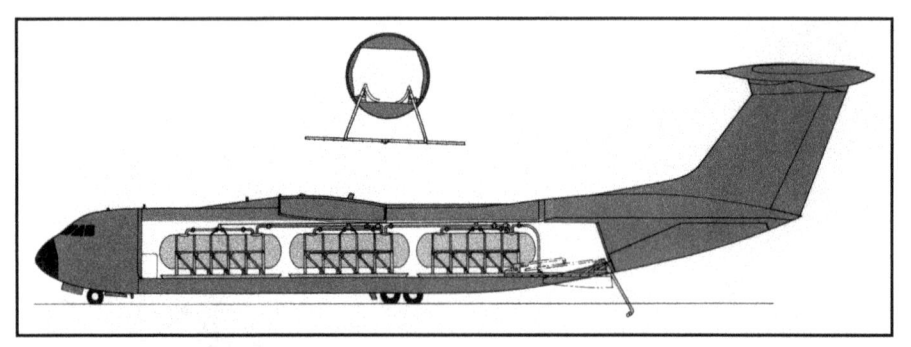

In an era when the U.S. Air Force used area defoliation in an effort to deny enemy ground forces the cover of vegetation, Lockheed suggested the C-141 could be a large-scale defoliant delivery aircraft. (Lockheed-Georgia via Craig Kaston)

heed had to reinforce about 155 square feet of fuselage structure. Special bearings and isolators used air cushions to float the telescope free from undue airframe contact.

In its new role, the civilian Starlifter inaugurated its telescope with observations of the comet Khoutek in 1973, followed by galactic discoveries including rings around Uranus, discovery of a heat source in Neptune, and the first direct observation of an atmosphere on Pluto.[38] Named the Kuiper Airborne Observatory (KAO), the Starlifter was a very active element of NASA's airborne science program out of Ames Research Center in California until the airplane's retirement there in the late 1990s.

What Might Have Been

Lockheed-Georgia committed to paper a number of C-141 spin-offs and additional roles that ultimately went unbuilt; variants as radical as bombers and air defense missile platforms. These proposed offshoots were characterized in a company brochure as having been "proved feasible by advanced design and development studies."[xxxix] Ideas for the Starlifter included:

Bomber: Triple bomb bays in the belly of a C-141A-size airframe were said to be capable of holding 54,000 pounds of bombs for accurate release from as high as 45,000 feet.

Missile Launcher: Sporting an enlarged radome, any of three air defense missile-launching Starlifter platforms would have revolutionized defense fighter concepts. One version would have used a forward missile bay to fire from, another would launch weapons from the back ramp, while the most ambitious makeover called for a rear-facing pneumatic tube exiting a modified aft fuselage with lengthened tail cone.

Airborne Warning and Control System: Known forever by its acronym, the Air Force AWACS mission ultimately went to a Boeing 707-based airframe. Lockheed suggested the job could be done by a special C-141 variant mounting a large dorsal saucer-shaped radome on pylons above the wing.

Missile Range Tracker: With a bulbous radome less pronounced than those installed on Boeing C-18 and EC-135 tracker aircraft, the proposed Starlifter variant "contains equipment required for spacecraft tracking and communications to serve as an aerial relay and control station for world-wide deployment in support of Apollo/MOL missions," according to the company.

Navigation Trainer: Lockheed-Georgia suggested a long-radome C-141 navigator trainer fitted with rows of individual navigator training stations for Starlifter as well as C-5 navigators.

Air Search and Rescue: With an extendable V-shaped line catcher inspired by the devices used on HC-130Hs to pluck downed fliers from the ground, this would have been an electronics-equipped rescue jet with greater range and speed for the mission.

Tanker: The use of C-for-Cargo airframes as refueling tankers was established and prudent; Lockheed hoped to garner a share of that market with Starlifter tanker derivatives on paper. Using either probe and drogue refueling or a flying boom mounted aft of the pressure bulkhead, with the boomer located in the pressurized compartment, some of the variants were drawn to permit them to take

Exaggerated by telephoto photography, a batch of C-141As seems dwarfed by a C-5A at Rhein-Main Air Base, Germany. Unlike C-141s, the C-5 uses wing leading edge devices to enhance low-speed flight characteristics. (Air Force photo via Don Macaluso)

Side by side, a C-141 Starlifter (50227) and a C-5 Galaxy (80228) show construction differences that can be used as recognition features to distinguish the two types of aircraft when size comparisons are not possible. The C-141 vertical fin tapers in at the top; the C-5 vertical leading and trailing edges are essentially parallel. The C-5 has a larger tail cone behind the bottom of the rudder.

fuel from the aircraft's own integral wing tanks, as well as from fuselage-mounted tanks. The flying boom version with the largest fuselage tanks was said to be able to carry 66,920 pounds of fuel on a mission with a radius of 2,370 nautical miles.

Air Weather Service: With an internal fuselage fuel tank plus equipment and stations for meteorology tasks, a proposed Air Weather Service C-141 was said to have an endurance of 17-1/2 hours without refueling.

Air Photographic and Charting Service: C-130s already performed this mission; Lockheed touted the Starlifter's increased range, speed, and capacity as reasons to buy Starlifters for the role.

Communications Platform: With limited modifications, Lockheed said, the C-141A could be converted to a high-performance communications platform. Ten communications personnel could be accommodated in mobile mission trailers loaded aboard the C-141. Lockheed figures showed a 10.7-hour loiter time on a mission with a 500-nautical-mile radius.

TAC Airborne Command Post: Lockheed looked ahead to fill a gap until the advent of AWACS aircraft in the 1970s by proposing a C-141 TAC

The general success and acceptance of the effective Boeing (McDonnell-Douglas) C-17 Globemaster III owes a lot to the pioneering notions of global airlift that germinated with the C-141 in the 1960s. Ironically, just as the C-141 was conceived as a convenient amalgamation of existing technologies, so was the C-17 touted as a recipe of off-the-shelf ideas and hardware in its era. Though the C-17 underwent extensive developmental flight testing to get into service, it also used an incremental capabilities growth program similar to the C-141 envelope expansion that led to ever-increasing operational utility. (Photo by Frederick A. Johnsen)

Holding its four engines close to the fuselage like the C-141, the Ilyushin IL-76, if not a copy, must at least be considered an homage *to the same airlift sensibilities that informed the design of the Starlifter. This example made a stateside visit in July 1992 (in the post-Soviet era).*

airborne command post in the interim. The company suggested the command post could be built with integral stations for crew and equipment, or a standard C-141 could be converted by installing vans. The inclusion of in-flight refueling receiver capability was touted as an option to greatly extend the aircraft's time on station.

Minelayer: Lockheed-Georgia suggested the C-141 could carry 44,500 pounds of mines on a 2,200-nautical-mile-radius mission, 50 nautical miles of which would be flown near sea level for saturation mine laying.

Defoliage: Lockheed-Georgia sketched a C-141A carrying three large internal tanks capable of delivering defoliant chemicals dispersed through the aft fuselage, citing the Starlifter's range and speed as benefits over C-123 and Huey helicopter defoliation aircraft used in the 1960s.

Flare Launcher: Before the maturation of night vision and other all-weather targeting devices, parachute-borne flares still had a place in 1960s combat operations. Lockheed suggested the C-141 could carry as many as 768 flares, or could loiter at altitude for as long as 12 hours, spanning the night. One artist's rendering shows flares exiting crew doors ahead of and behind the left main landing gear pod.

Tactical Support: To accommodate substandard tactical airfields, Lockheed suggested a tactical support version of the C-141 using special six-wheel main landing gear to spread the weight over the ground. Design payload of this variant would have been 63,255 pounds, according to a company presentation.

Electronics Platform: Another big-radome suggestion was a generalized electronics platform C-141 for unspecified special mission requirements.

While most of the company's creative ideas for C-141 special missions remained paper airplanes, at least one of Lockheed-Georgia's suggested uses for Starlifters was embraced by the Air Force:

Special Purpose Research and Development: The company literature said: "In terms of volume and overall performance characteristics, the C-141 is unmatched among present-day aircraft for a myriad of special-purpose research and development missions calling for a modern jet airplane with space and service facilities for special installed equipment."[40] Ultimately, the early C-141s ordered in fiscal year 1961 that figured prominently in the Starlifter developmental test program were retained by the Air Force as test-bed aircraft based at Wright-Patterson Air Force Base, Ohio, until they were assigned to the Air Force Flight Test Center's 418th Flight Test Squadron at Edwards Air Force Base, California in 1993. In this role, these Starlifters were the last A-models in service, designated NC-141A.

Lockheed promoted the notion of Starlifters as Airborne Warning and Control System (AWACS) host airframes in an artist's rendering from the 1960s that suggests the presence of an aerial refueling receptacle. (Lockheed-Georgia via Craig Kaston collection)

Japan's twin-engine Kawasaki C-1 embraced the circular fuselage, podded main landing gear, and high-mounted wing that made the Starlifter so versatile, though it all came in a smaller package. (Photo by Frederick A. Johnsen)

T-Tailed Counterparts

The advent of the C-141A revolutionized airlift for the U.S. Air Force, and its general features can also be found in other jet transports around the world. Even before the first Starlifter flew, an industry observer in 1963 compared the wing of a proposed swept-wing, T-tailed, jet-powered version of the Short Belfast transport to that designed for the C-141.[41] Though the Belfast jet variant did not materialize, other look-alikes followed the C-141 into the skies.

Japan's twin-engine Kawasaki C-1 embraced the circular fuselage, podded main landing gear, and high-mounted wing that made the Starlifter so versatile. Work on the Japanese design began three years after the rollout of the first C-141A. The first prototype flew on November 12, 1970. Testing and trials lasted until the spring of 1973; production C-1s entered the Japanese Self Defense Force inventory in December 1974. Production rates were low by U.S. standards – about 36 aircraft in the original configuration. Various portions of the C-1 were built by several Japanese aerospace manufacturers. As expected, this medium-range twin is smaller than the C-141A, with the C-1's wing spanning 100 feet, 5 inches compared to the Starlifter's 160 feet, 1 inch.

Nearly eight years after the first flight of the C-141A, in 1971, the Soviet Union introduced the Ilyushin Il-76T, a high-wing four-jet transport bearing an intriguing resemblance to the C-141A, with typical Soviet touches including substantial glazing in the lower nose section. The Il-76 is only a few feet larger in every dimension than a C-141A, and its Soloviev D-30KP turbofans are said to produce more than 26,000 pounds of thrust apiece. The Il-76 has been used by the Soviet air force as well as by the Soviet airline Aeroflot.

The sincerest form of flattery for the Starlifter design must be Lockheed's own C-5A Galaxy, winner of the 1965 U.S. Air Force heavy lift CX-HLS competition. When size comparisons are not at hand, C-5s are occasionally mistaken for the smaller C-141s. Quick recognition features distinguishing the two Lockheed jet transports at a glance include the use of rounded wing and horizontal stabilizer tips on the C-5, and the appearance of a constant-chord vertical fin and rudder on the C-5 instead of the tapered assembly on the C-141. And yet in early 1965, a Lockheed artist's rendering of its CX-HLS proposal showed a beefy transport with a tapered vertical fin and squared wing and stabilizer tips much like those of its smaller C-141 sibling, barely two months before the Air Force was to receive industry proposals for the impending C-5A.

Jet Transport Comparison Data

Model	Length	Span
C-141A	145'	160'
C-5A	247'10"	222'9"
C-1	95'2"	100'5"
Il-76T	152'11"	165'8"

Model	Height	Weight (Gross)
C-141A	39'3"	325,000 lbs
C-5A	65'2"	769,000 lbs
C-1	32'9"	99,210 lbs
Il-76T	48'5"	374,785 lbs

Model	Thrust (Per Engine)	Range
C-141A	21,000 lbs	3,675 nm
C-5A	41,000 lbs	3,749 mi.
C-1	14,500 lbs	2,084 mi.
Il-76T	26,455 lbs	3,100 mi.

Statistics compiled from several sources; some variation may occur. Range estimates depend on payload versus fuel quantity trade-offs.

GETTING THE JOB DONE 4

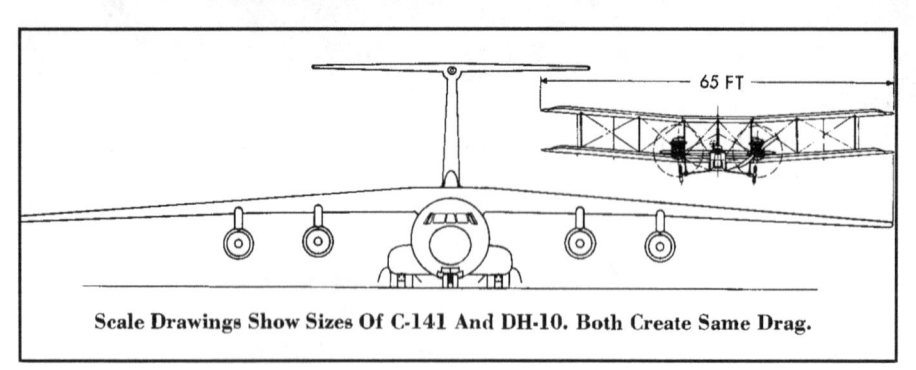

Scale Drawings Show Sizes Of C-141 And DH-10. Both Create Same Drag.

Lockheed prepared a drawing of the frontal aspects of the C-141A and a 1920s vintage DeHavilland DH-10 twin-engine biplane. Both aircraft, despite great differences in size, were estimated to produce the same amount of drag. (Lockheed-Georgia)

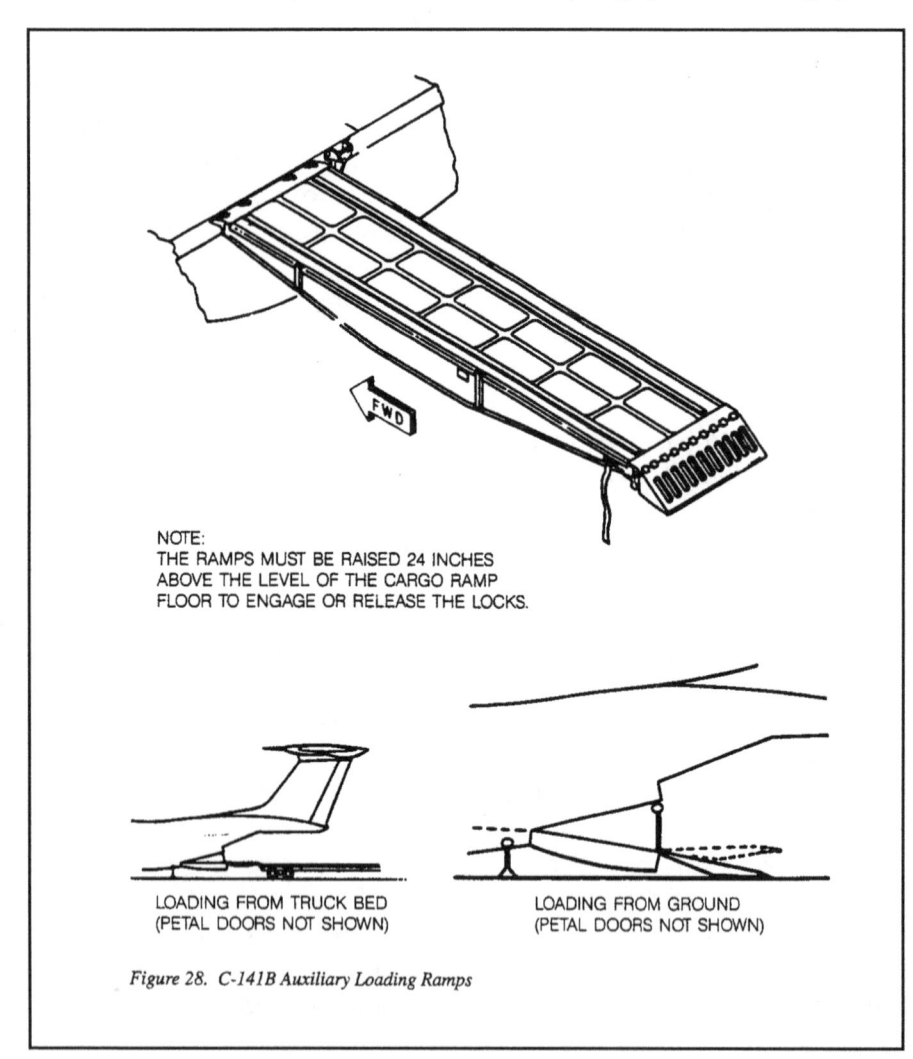

NOTE:
THE RAMPS MUST BE RAISED 24 INCHES ABOVE THE LEVEL OF THE CARGO RAMP FLOOR TO ENGAGE OR RELEASE THE LOCKS.

LOADING FROM TRUCK BED (PETAL DOORS NOT SHOWN)

LOADING FROM GROUND (PETAL DOORS NOT SHOWN)

Figure 28. C-141B Auxiliary Loading Ramps

Auxiliary load ramps could be attached to the hinged loading ramp of the C-141 to enable vehicles and personnel to enter and exit from ground level. (Lockheed Martin)

Starlifter Support Equipment

The C-141 was made to be compatible with the Air Force's 463L loading system. Long open-bed vehicles with small cabs offset to the side can deliver pallets at deck height to the C-141. Roller beds on these trucks and removable rollers in the Starlifter's floor enable quick manual loading and unloading of standard Air Force aluminum cargo pallets. Standard aerospace ground equipment (AGE) – maintenance stands, power carts, and heaters are serviceable for Starlifter operations. Specialized AGE was created to allow component testing by MAC units for use with the C-141's computerized All Weather Landing System (AWLS).

Some accommodations were made at bases receiving C-141As in the 1960s. Hangar doors required wide cutouts in the top center portion to permit passage of the high T-tail through openings formerly reserved for conventional-tailed aircraft like C-124s.

"Lockheed Fixes" and Tricks of the Trade

Even the reliable Starlifter can develop bugs during a mission – problems requiring maintenance attention at the next stop. Squawk sheets listing faulty items are pored over by mechanics who tackle the problems. Sometimes, airborne maladies disappear once a C-141 lands; the mechanics' answer on a squawk sheet would typically note "Could not duplicate on ground." A Starlifter that seemingly healed itself was said to have performed a "Lockheed fix."

Some problems involve aerodynamics and can only be experienced, and the fix validated, in flight. The potential for out-of-rigging problems with operating the petal doors, pressure bulkhead, and loading ramp for airdrops sometimes prompted aircrews to cycle these components shortly after takeoff to ensure proper operation. If the parts failed to align correctly during cycling, a return to base was in order to fix rigging on the ground, an arcane art since the static loads on these parts differ from flight loads.

With the advent of aerial refueling capability on C-141Bs, it was prudent for flight crews to open the refueling receptacle doors in flight shortly after takeoff, especially when departing a wet locale, to air dry the hardware to prevent freezing at altitude that could thwart a refueling.

The streamlined main landing gear pods of the C-141 create a cavity that aircrews have used for cold storage on flights. Perishable food items have been placed in the pods where they remain in cold storage at altitude, ready to be retrieved upon landing. The gear pods evolved in use; an early intent was to stow life rafts in a compartment at the rear of the faired left gear pod. Subsequently, the left gear pod compartments were used to stow engine inlet covers. On C-141As, the gear pod stowage compartment was easily accessible by opening the aft fuselage emergency exit on the side of the fuselage. But with the insertion of the aft fuselage plug during B-model elongation, the emergency hatch was effectively moved back 10 feet.

Before the routing of aerial refueling plumbing on B-models blocked its location, C-141As featured a small port in the top of the cabin for inserting a periscopic sextant. This feature invited a number of ingenious uses, recalled C-141 pilot Keith Littlefield. Dutch roll recovery training, in which a C-141A experienced radical yawing while pilots learned how to avoid the perils of Dutch roll that nearly claimed a Starlifter over Vancouver Island, prompted the pilots to install the periscopic sextant and train it on the Starlifter's tail. The tail could be seen to bend in the forces it encountered during these maneuvers, Littlefield recalled. There was also a way, Starlifter crews learned, to place a hard-boiled egg in the sextant opening and use differential air pressure

An Air Force K-Loader vehicle removed a pallet of cargo from a C-141 supporting Operation Enduring Freedom in Romania. (U.S. Air Force photo by MSgt. Jon Nicolussi)

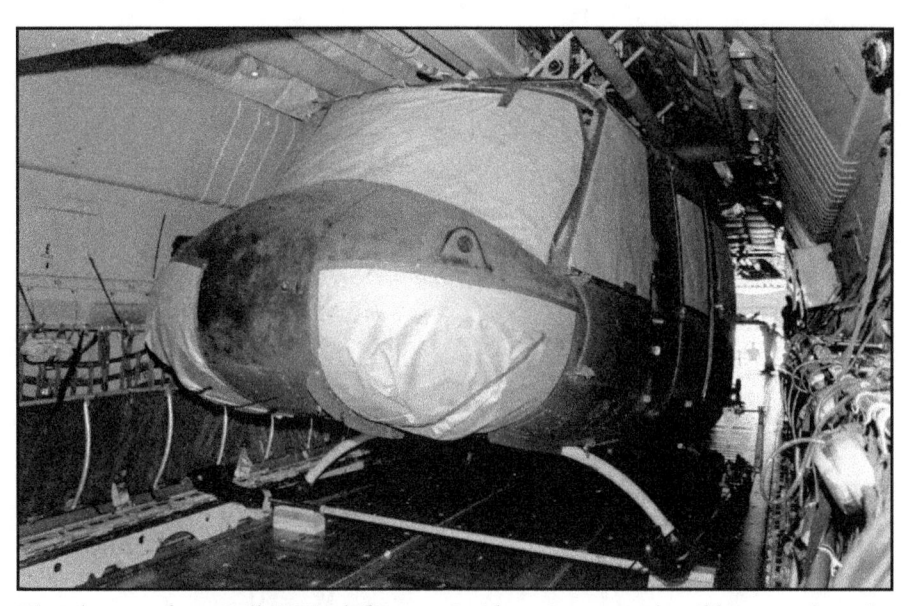

There's room for a Bell UH-1 helicopter inside a C-141, with cribbing to place the helicopter in the right position.

to launch it in flight. Littlefield recalled that C-141A aircrews were admonished to discontinue this practice to preclude a build-up of egg remnants in the sextant port to the annoyance of navigators.[42]

After a long journey in the MAC system, it was common for aircrews to do a little housekeeping on their Starlifter before returning to home station. In the C-141A, the natural vacuum created by the pressure differential between the cabin and the

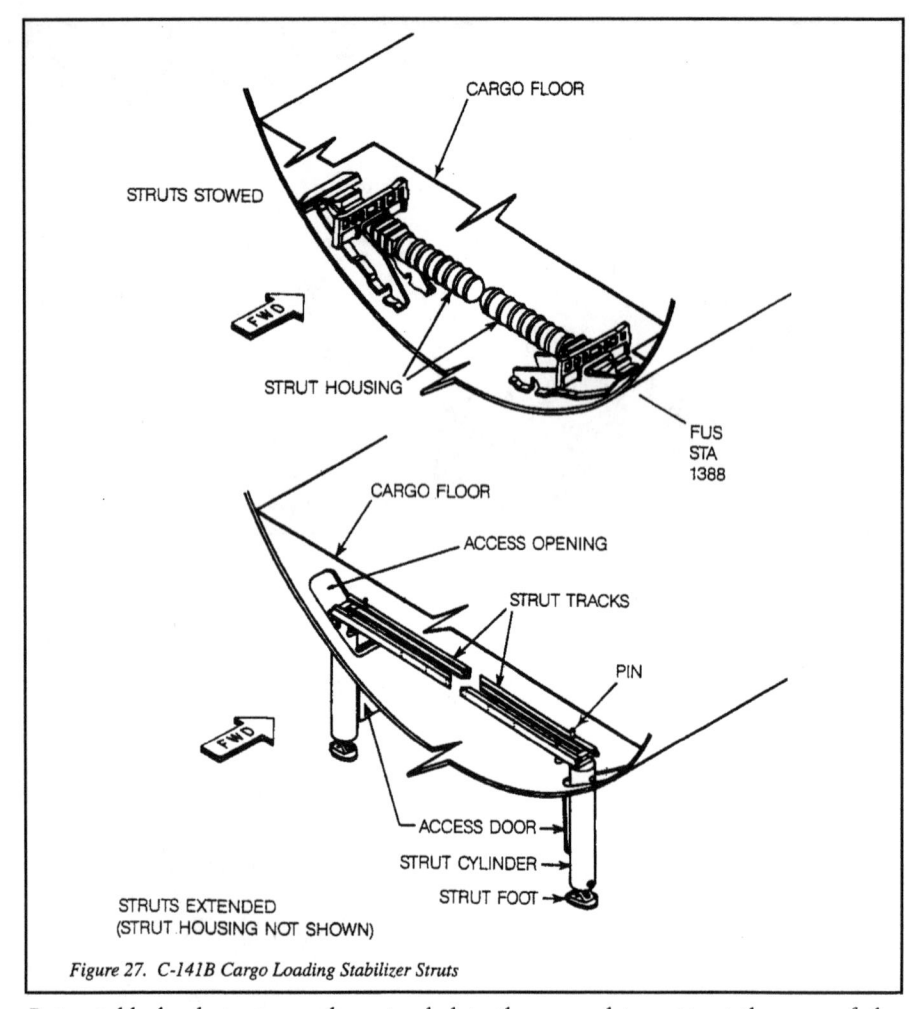

Figure 27. C-141B Cargo Loading Stabilizer Struts

Retractable load struts can be extended to the ground to support the rear of the C-141 as heavy cargo weight rolls on board from loaders. (Lockheed Martin)

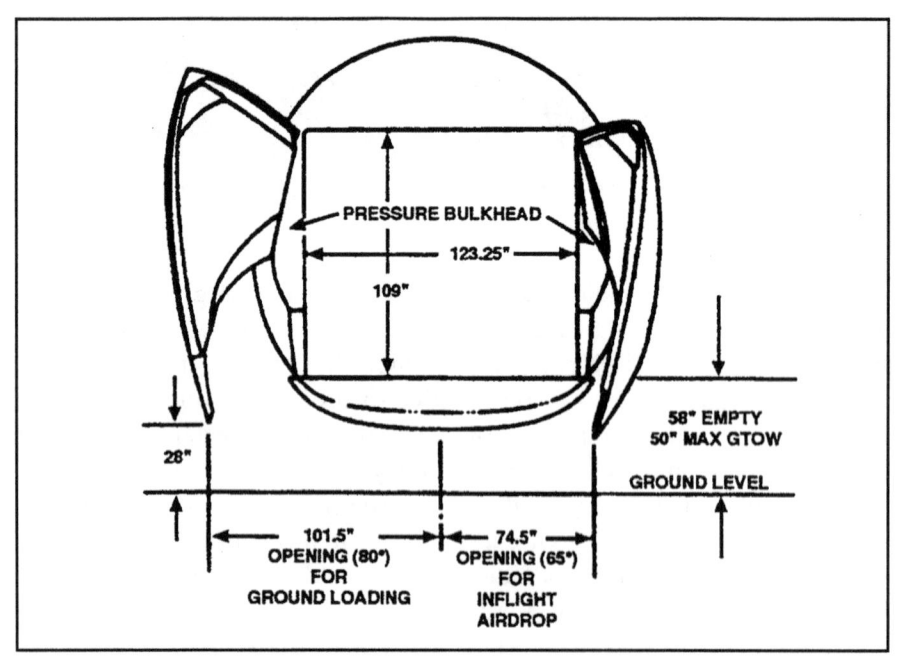

Starlifter petal doors can be opened wider for ground-loading clearance than for in-flight airdrop activity. (Lockheed Martin)

outside air led to an ingenious hook-up of a vacuum cleaner hose to the overhead sextant port, Littlefield said. While this worked wonders in tidying up the flight deck, it could also spew small debris like loose screws back against the vertical tail where impact damage could occur. Littlefield recalled that a screen installed in the system was intended to preclude this type of foreign object damage.[43]

Problems and Solutions

The C-141A's fast track into service meant test C-141As were still in evaluation at the Air Force Flight Test Center while younger A-models were entering the MATS (later MAC) airlift stream. This concurrent test and production schedule was characterized as placing Starlifters in service two years earlier than traditional sequenced procurement programs, although that two years may have been filled with attention to fixes and temporary limitations on aspects of the C-141A's flight envelope.

In an effort to predict where problems were likeliest to occur as the C-141A fleet aged, a Lead the Force program was instituted following discussions in the fall of 1964. Select C-141As were to be flown more hours than the rest of the Starlifter fleet, eventually representing a use rate two years in front of the main force of Starlifters. Lead the Force programs had their beginnings in Strategic Air Command as early as 1959, when Gen. Curtis LeMay wanted the ability to forecast future problems with his fleet of nuclear bombers and supporting tankers.

Lead the Force aircraft were intended as predictors of structural fatigue and statistical data about fleet service life. One Air Force planner said Lead the Force was a way to answer "Where will the airplane fail next?" By observing the

A series of latches helps keep the C-141 pressure bulkhead secured during flight at altitude. This example was photographed in the Travis Air Force Museum's C-141B. (Travis Air Force Museum collection)

With the Starlifter's petal doors in the ground-loading position (open wider than the airdrop mode), a U.S. Army UH-1 helicopter, minus rotor blades, is eased inside, circa August 1982.

faster-aging Lead the Force aircraft, trends might be detected that would give the C-141 team time to devise a fix before a problem became a fleet-wide issue. To maintain the realism and totality of Lead the Force, when engines were removed from these Starlifters for maintenance, they were to be returned to a Lead the Force C-141A "at the earliest opportunity," according to a program plan draft. The C-141 Lead the Force program officially began on the first of June 1965. By October 1968 it was concluded, with the stated flying hour and total landings goals achieved.

Pressure Bulkhead Failures

Though the Starlifter program overall delivered a remarkable and serviceable transport that vaulted USAF airlift to the forefront in military planning and execution, its most dramatic early problem was a series of pressure bulkhead failures in flight, sometimes tearing off one or both petal doors in the process.

Early-morning sun illuminates K-loader forklift vehicles placing cargo pallets in C-141s during a 1982 exercise. Loading struts are extended on the Starlifters to support the weight of the pallets as they pass from the loader to the aircraft.

To create a pressurized cargo compartment and still permit drive-on loading of large loads and airdrops of big items, it was necessary to devise a large rectangular pressure bulkhead that was hinged at the top and swung aft to clear the opening for unpressurized operations. Some structure deficiencies were caught and corrected during static tests in 1963 and 1965.

But the corrections did not prevent a dramatic blowout on July 10, 1966, as a C-141A (65-0220) of the 437th MAW was climbing through 19,000

A cherry-picker boom truck enabled a Starlifter maintainer at McChord Air Force Base to reach the top of the high wing to spray de-icing fluid before the C-141A flew, circa 1981.

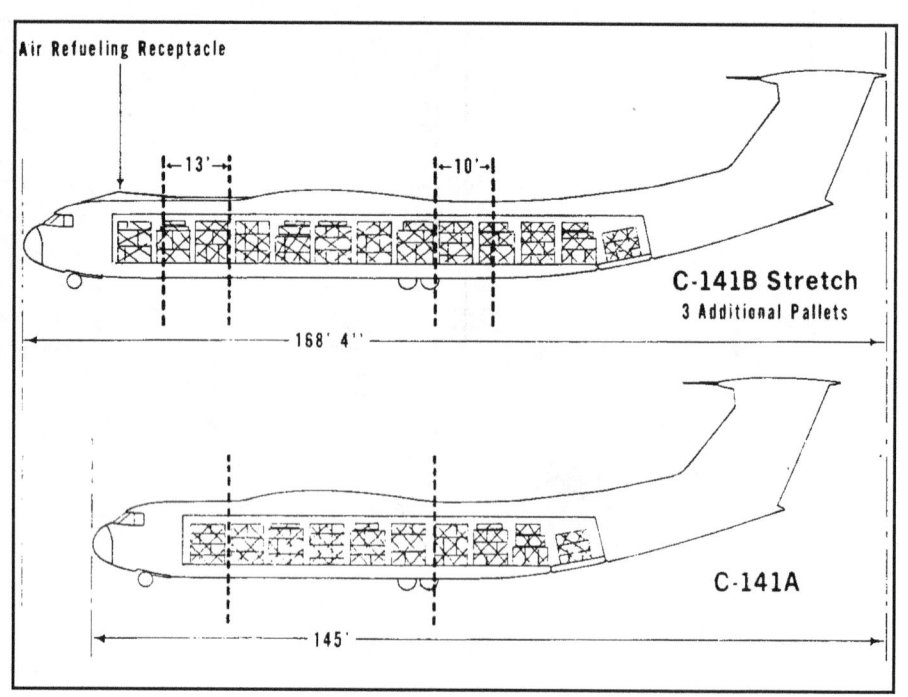

feet. The pressure door ramp latches, securing the bulkhead to the Starlifter's floor, failed, allowing the bulkhead to rotate aft, striking the overhead loft area of the fuselage. In rapid succession, all petal door hinge fittings broke from the longerons to which they were mounted, and the petal doors fell to earth. Subsequently, pressure door hooks were inspected and aligned as needed. Other Time Compliance Technical Orders (TCTOs) dealt with related pressure door inspection and modification procedures, which were also incorporated on aircraft still under construction, beginning with C-141A number 66-126 (delivered to the Air Force in November 1966).

The issue came to the fore again on January 15, 1967, when C-141A 65-0230 suffered an explosive decompression while climbing to altitude a half hour after departing Wake Island with 96 troops bound for Southeast Asia. Again, both petal doors tore free from the Starlifter. The rapid loss of cabin pressure, accompanied by the lifting of the huge pressure door, could result in an instantaneous tempest of blowing dust and debris, and possibly a short-duration fog as a barometric result of the rapid air pressure drop.

Concern that such events could suck troops into the slipstream led to the installation, for awhile, of a net in front of the pressure bulkhead. Airmen, employing the time-honored military pastime of nicknaming, came to call the net a "people strainer," remembers former C-141 crew chief Ben Howser, whose 62nd MAW C-141A was one of the aircraft beset with a pressure door blowout. In the aftermath of the Wake Island incident, passengers were required to keep their seatbelts buckled. At this time, the allowable pressure differential was lowered from the previous 8.6 pounds per square inch (PSI) to no more than 7.3 PSI. Missions transporting personnel were limited to a maximum altitude of 33,000 feet; cargo missions were not to exceed 39,000 feet, and aeromedical evacuation flights were to be kept at or below 29,000 feet.[44]

Reviews of the pressure bulkhead and its latching system led to efforts by Lockheed and the Air Force to

The economy of stretched Starlifters was graphically depicted in a line drawing used by the Air Force. (WR-ALC)

When viewed from the front, the petal doors extend well outside the diameter of the C-141's fuselage during ground loading; airdrop opening is less wide.

design both interim and final fixes to the problem. The 437th MAW, whose aircraft had experienced the first in-flight C-141 pressure door release, designed cam jacks to hold the pressure bulkhead in position with hooks engaged until the aircraft was pressurized. Another mechanical fix used seven links mounted to the forward (cabin) side of the pressure door, which were engaged with the aft row of cargo ramp tie-down fittings. These devices provided a secondary latch system to restrain the door if the primary latches gave way under pressure. Researchers at the Air Force's Warner Robins Air Materiel Area noticed the pressure door's pliable seal could fold back improperly under the door, which could keep the door from being latched properly.

Ultimately, a series of fixes, including evolved pressure-door restraints and improvements that stopped a series of early honeycomb delamination problems in pressure doors, as well as experience with the aircraft in service conditions, rendered pressure door blowouts a thing of the past.

Weight Watcher

The Starlifter's creators at Lockheed-Georgia kept close tabs on the weight of their new jet. Assembly-line workers were admonished to do the job right; "too many patches or fixes can add up to hundreds of pounds by the time an airplane rolls out the B-1 (assembly building) door," a story in the Lockheed Southern Star employee newspaper said. The story cited one particular bolt throughout the C-141, which, if exchanged for another that was one size too long (taking into account the extra washer per bolt that would be required), would result in an extra 250 pounds for the aircraft to lug around. The article's author said: "Anybody in Manufacturing can tell you that C-141 weight control is demanding. It calls for the greatest know-how and ingenuity ever asked of us. Achieving the necessary strength-per-pound is a task requiring the absolute utmost in material and craftsmanship excellence." The article, written in January 1963, months before the first rollout, said the C-141 was just under its required empty weight at that time. Exceeding that weight could result in a stiff penalty

Airlift details, January 1982: An interior C-141A view looking aft shows the fuselage ringed with insulating panels and storage compartments. Near the front, rear-facing airline-style seats, sometimes called "Boeing seats" by aircrews, have been attached to the floor. Farther aft, red nylon troop seats are stowed folded along the fuselage sides so cargo pallets will fit. The loadmaster is walking aft between rows of rollers that can accommodate pallets. When a flat floor is required, strips of rollers can be removed, turned over, and installed flush with the floor. Circular lighting runs along the centerline of the fuselage ceiling. At the far right, just above the Boeing seat, the oblong container is one of many passenger oxygen mask containers hooked up to the Starlifter's oxygen system, ready in the event of depressurization.

Using vehicles at ramp height, cargo pallets and airdrop loads can be rolled straight in the back of a C-141, braced by its deployed load struts.

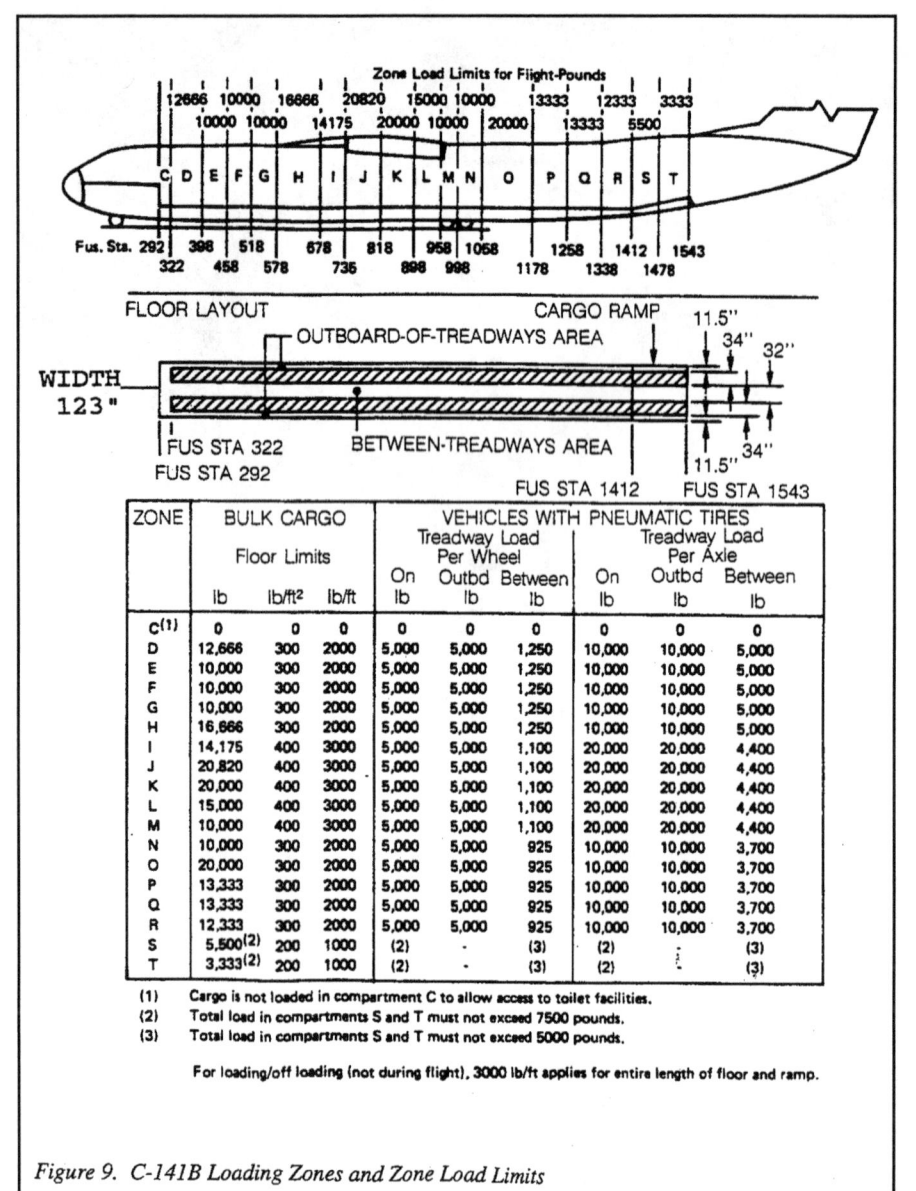

Figure 9. C-141B Loading Zones and Zone Load Limits

C-141B loadmasters could refer to this chart showing vehicle and cargo load specifications. Treadway areas could support more pneumatic-tire vehicle weight than could the flooring between the designated treadways. (Lockheed Martin)

Bases hosting C-141s needed infrastructure including movable stands to service the Starlifters' tall tails. (Travis Air Force Museum collection)

A cargo pallet inside a Starlifter is heavily wrapped in stout plastic to keep rain out when the cargo is marshaled. Webbing helps keep the boxed cargo in place. The metal cargo pallet rolls on four strips of rollers on the C-141's floor that can be flipped over to provide a flush, rollerless surface if needed. Cargo locks will hold the pallet in place.

from the Air Force. "We push to avoid overdesigning and overbuilding," the author said.[45]

Vortex Generators Removed

Initially C-141As had vortex generators – small metal blades attached to the upper surface of portions of the wing. Corrosion was an issue where the vortex generators attached to the wings. Additionally, Lockheed tests showed that a C-141A with vortex generators removed attained a small improvement in cruise performance. An order went out for the vortex generators to be removed in the field.

APUs Changed to Vent Sideways

The auxiliary power unit (APU) for the C-141 is carried in the left main landing gear pod. As built, the APU's exhaust vented out the top of the gear pod, resulting in an accumulation of hot exhaust residue on the underside of the wing root. A change resulted in the rerouting of C-141 APU exhaust out the side of the gear pod to alleviate this problem.[46]

Depleted Uranium Counterbalances

Aircraft control surfaces require balance to act properly; mass balances are heavy weights that facilitate

When it becomes necessary to do a landing-gear retraction test in the field, jack points on the Starlifter's fuselage and under the wings allow the entire aircraft to be elevated to free the wheels.

At a November 1982 open house, a C-141B awaits visitors with its ramp lowered to the tarmac and auxiliary loading ramps installed to facilitate walk-on, walk-off foot traffic. Yellow markings outlining emergency exits were subsequently toned down on European One camouflaged Starlifters. (Photo by Frederick A. Johnsen)

proper control movements and forces. On the C-141, the ailerons and elevators were mass balanced with heavy depleted uranium. As a result, during aileron removal and replacement maintenance tasks, Starlifter maintainers wore dosimeters to indicate radioactivity levels.[47]

B-Model Tail Scrapes

When C-141Bs entered service, with the fuselage behind the main landing gear now 10 feet longer than before, the possibility of the lengthened aft fuselage contacting the runway during landing was increased. Several tail-scrape incidents were documented, and MAC flying safety offi-

A fuel truck pumps JP-4 jet fuel into a C-141A at Dover Air Force Base, Delaware. Receptacles are located in the landing gear pod. This view shows a portion of the main landing gear that protrudes above the contours of the streamlined gear pod when extended, necessitating a door in the top of the pod to accommodate it. (Air Mobility Command Museum collection via Harry Heist)

Special runs of skate rollers were arranged in a C-141B for midwinter Antarctic airdrop resupply missions. The pressure bulkhead is closed and latched in the photo; short-hinged extensions to the skate rollers can be flipped open when the bulkhead is raised. Airdrop cargo bundles could be dropped off the rear of the ramp over McMurdo Sound at the edge of Antarctica. For airdrops inland, over the South Pole, the petal doors were left closed and bundles were rolled to the flush troop door openings. Crews were concerned that a malfunction of opened petal doors deep over Antarctica could hamper the Starlifter's ability to reach a runway to land. Dangling hooks above the latch restraints are for stowing the latches during bulkhead movement.

The pressure bulkhead inside a retired C-141B was photographed with the restraining latches stowed out of the way, using hooks installed for that purpose. Resting on the ramp are four auxiliary loading ramps. (Travis Air Force Museum collection)

cers made suggestions for changes in pilot technique to mitigate this problem. One correction was to add about 10 knots to the no-flap landing speed of the C-141B, recalled pilot Keith Littlefield.

Bleed Air Rain Removal and Early Anti-Icing Fluid

The C-141 used bleed air pressure from the engines to force air over the front windscreen panels as a means of rain removal instead of using mechanical wipers. Some crews were in the habit of turning on the rain removal air pressure on the ground before a flight on a rainy day to blow out any accumulated moisture, lest it freeze at altitude and possibly inhibit in-flight use of the rain removal feature. When selected on the ground, the rain removal system could produce a visible spray over the cockpit as the forced air ejected pooled water.[48]

Good Reputation

If the deliberate concurrent procurement and development of the C-141 spotlighted some issues early in the Starlifter's career, rapid introduction of Starlifters into service paid its own dividends. Military Airlift Command historians analyzing the C-141 program in 1973 reported: "...After Fiscal Year 1968, very little can be found in the MAC histories concerning C-141 system achievements or failures. While in previous MAC histories many deficiencies are recounted, they are always tempered with facts indicating that the C-141 did indeed perform well. In fact, it can be safely stated that MAC's contribution to the war effort and other Department of Defense airlift tasks during the 1965-1971 period could not have been achieved without the rapid acquisition of the C-141 Starlifter."[49]

An early photo of a MATS C-141A at Dover AFB, Delaware with a staged cargo load gives a reference to the bulk that can be handled. But stanchions leading up the loading ramp suggest this is a display, not a working load, about to be placed aboard the Starlifter. (Air Mobility Command Museum collection)

The design of Starlifter petal doors had to permit more than simply fairing the aft fuselage. The doors needed to get out of the way of ground loading, and, when airdrop was added to the list of C-141 tasks, a redesign was needed to permit them to open partially to accommodate that job. An early concept had the doors hinging outward along the forward edge of the doors – suitable for ground use, but more in keeping with huge speed brakes for airdrop. The production version accommodated both airdrop and airland needs.

A Bell OH-58 helicopter is eased into a C-141 circa August 1982 by handlers using planks placed on the auxiliary loading ramps and a snubber strap to control the helicopter's fuselage position.

PAINT AND MARKINGS 5

Shiny new MAC C-141A 64-0622 showed variations in natural-metal skin panels and painted silver-gray areas on the tail and top of the wing. In service, natural metal C-141As increasingly had sections of metal covered with paint for corrosion control. Ultimately, the fleet was painted gray and white overall. (Marty Isham collection)

After the crew of C-141B 65-0258 airdropped a load near an incorrectly parked truck during an airdrop competition, the aircraft was decorated with temporary markers including a truck "kill" symbol and the inscription "62 Bomb Wing." (62 MAW/PA)

Silver "speed tape" made impromptu temporary names possible for Starlifter crews – and the Volant Rodeo airdrop competition aircraft involved in the truck incident received its due. (62 MAW/PA photo by Shirley Snavely)

Natural Metal Finish

The C-141 entered Air Force inventory in the waning years of uncamouflaged warplanes. The test fleet C-141As, as well as production A-models to follow, were delivered in shiny natural metal finish with prominent national insignia. When corrosion around some fuselage rivet lines became an issue, the natural metal C-141As began to take on a "belted" appearance as silver/gray paint was applied in narrow bands along some rivet lines around the circumference of the fuselage. The lower nose section and vertical fins of silver C-141As were also painted gray.

The top left wing accommodated a bright red, white, and blue 50-inch star insignia, and the top right wing carried a 50-inch-high USAF in insignia blue (Federal Standard 15044). The upper wing markings were parallel to the aft walkway border marking, and hence followed the sweep of the wings. These markings were repeated on the opposing lower wing surfaces. Bright 50-inch stars and bars also adorned both sides of the aft fuselage. A smaller "U.S. AIR FORCE" in 21-inch-high letters was applied to both sides of the forward fuselage using insignia blue (color FS15044). Radomes were to be black (FS17038).

Gray and White Scheme

By the early 1970s, C-141As were painted two-tone white (FS17875) over gray (FS16473), an innocuous scheme that provided a measure of corrosion control, but didn't look as militaristic as camouflage. An early iteration of this color scheme used on a few Starlifters relied on paints that had an unfortunate affinity for dirt and stains; the fleet later switched to polyurethane paints that withstood grime better. In later years when most C-141Bs were given European One camouflage colors, a select few Starlifters retained the gray-and-white paint, and were tapped for mis-

52

sions to foreign locations where local sensitivities might be agitated by the sight of an American military aircraft in full tactical camouflage.

A three-inch insignia blue (FS15044) horizontal stripe on the fuselage delineated the white top and gray lower fuselage colors. The stripe was bisected by the fuselage star insignia; the forward fuselage "U.S. AIR FORCE" markings were 13 inches above the stripe.

European One Camouflage

The 1980s saw C-141s – mostly B-models, but a few pre-conversion C-141As as well – given variegated green and gray European One camouflage paint. The paint was selected to give Starlifters some protection in the event of military action in European type environments. Colors (with Federal Standard numbers) included green (FS34092), green (FS34102), and gray (FS36118). The gray is a close match to the World War II neutral gray used on the undersurfaces of camouflaged U.S. Army Air Forces aircraft. Variations in pattern and even some color substitutions on some C-141Bs occurred over time. Radomes were to match the adjacent fuselage color (green FS34092). Undersurfaces, including the lower surfaces of the wings and horizontal tail, and the lower surface of the fuselage, were gray FS36118.

Initially, the European One camouflage extended over the lip of the engine cowl inlet. Since this metal lip was used to transmit hot air from the engine to keep ice from forming on the inlet, the camouflage paint soon flaked away from the lip in the pattern of the heat sources, leaving spotty paint coverage. Later, it was typical for European One C-141Bs to have bare metal cowl inlets. Although a

The C-141A's natural metal finish was followed by an almost airliner-style white-topped gray paint scheme through the 1970s and on a few MAC C-141Bs well into the 1980s. This photo of a 62 MAW C-141A (66-0158) dates to circa October 1981. White upper surfaces helped reflect heat, while the gray undersides provided corrosion protection and helped mask the appearance of some grime on the airframe between washings.

C-141As used for the return of POWs from Hanoi to the U.S. were emblazoned with red cross symbols on their vertical fins for easy identification. Aircraft 65-0258 flew with the 62nd Military Airlift Wing out of McChord AFB, Washington. (62MAW/PA)

C-141A 67-0021 featured an experimental shadowy gray camouflage scheme, photographed on a drizzly day in March 1979. (Photo by Keith Laird)

Decals noting increments of 100 flight hours were posted on Petunia, *C-141A 63-8075 used in the flight test program at Edwards AFB, circa 1965. (Travis Air Force Museum collection)*

AMC Proud Gray

European One green and gray camouflage was superceded by a uniform overall gray scheme with black outline lettering and numerals, prototyped on a C-141B and a C-130 in the summer of 1991. A bright exception to this was a horizontal band on the vertical fin labeling the base or the unit to which the C-141 belonged. Undertaken simultaneously with some interior upgrades to airlift aircraft, the overall gray paint was dubbed "AMC (Air Mobility Command) Proud Gray." The color matches FS26173 gray. The semi-gloss gray polyurethane simplified painting and was said to be beneficial in limiting corrosion. Its application saved time and money.

Experimental Paint Schemes and Variations

conscious effort was made to use appropriate camouflaged cowlings when swapping engines on C-141s, in some instances, European One Starlifters operated briefly with shiny gray engine nacelles intended for gray-and-white C-141s.

Some European One schemes had more than one paint color on the petal doors. This could cause irregularity if the petal doors had to be swapped for maintenance, and European One C-141s gravitated toward using solid gray covering the petal doors for uniformity.

Using subdued black markings on European One Starlifters did away with the large "U.S. AIR FORCE" legend on the forward fuselage sides, as many markings were downsized or deleted in the interest of camouflage integrity.

At least one C-141A sampled an experimental two-tone light gray camouflage, while a B-model temporarily carried patches of tan camouflage paint over its European One hide during a paint ablation test of new camouflage possibilities.

When the rapid dispatch to Saudi Arabia demanded all available C-141s for Operation Desert Shield in 1990, some Starlifters completing depot

The absence of a refueling hump atop the fuselage tags the first two Starlifters in this line-up as A-models. A few C-141As received the new European One green and gray camouflage paint before the aircraft entered stretch modification. (Air Force photo via 62MAW/PA)

maintenance were returned to their units after being stripped of paint, but before receiving new coats. This resulted in short-lived "silver bullet" C-141Bs flying vital missions in dulled bare-metal finish.

When C-141As were new at the 62nd Military Airlift Wing at McChord AFB, Washington, in the late 1960s, members of the Wing who were familiar with airliner radome paints used nearby at Boeing painted a number of the Wing's Starlifter radomes gray instead of the standard black. This nonconformity was not tolerated for long before instructions were issued to make McChord Starlifters match the rest of the fleet.[50] Photos from this era also depict white and two-tone radomes on C-141s from other units.

Markings

The Military Airlift Command's use of Associate Wings – Air Force Reserve wings aligned with an active-duty wing – provided additional crews to fly the C-141s assigned to the active-duty wings. It was common practice on non-camouflaged Starlifters to place the bright colored shields of the active duty and reserve wings side by side on the vertical fins of the Starlifters they flew. Black (or dark insignia blue) lettering on either side of the nose indicated the owning active-

This snapshot captured a C-141A with an extra anti-glare panel in black sweeping up from the radome in a style reminiscent of KC-135s. The dull silver overall sheen suggests this Starlifter may have been painted aluminum silver overall before the fleet settled on gray and white. (Stoney Burke via Marty Isham)

Photographed at Abbotsford, British Columbia in August 1968, C-141A 66-0145 of the 62nd Military Airlift Wing displayed its gray radome and its unit lettering on the landing gear door, which was painted white. Non-standard color variations like these were not long-lived. The ringed appearance of the forward fuselage is due to protective paint applied along some rivet lines where corrosion was found. (Photo by Frederick A. Johnsen)

ABOVE AND BELOW: *Some C-141As at Travis AFB sported non-standard white radomes in the 1960s. (Travis Air Force Museum collection)*

duty wing, as in "62nd MAW." In the global airlift system, it was common practice for C-141s to enter a pool; aircrews overseas might fly a Starlifter of the 63rd MAW on a 62nd MAW mission on an intratheater flight in the system.

The MAC shield was applied to both sides of the aft fuselage above the angled hinge line of the petal doors. This followed the earlier use of round MATS emblems, without the shield perimeter, on early C-141As.

Early in the Starlifter program, aircraft selected for accelerated aging were denoted with a large circular "LTF LEAD THE FORCE AIRCRAFT" sign beneath the cockpit.

With the advent of European One camouflage, all markings, including outline insignia, were black; use of MAC or airlift wing insignia was done in subdued tones instead of full color to keep camouflage integrity. Variations on black stencil markings continued with the advent of overall gray paint schemes on C-141Bs.

Throughout the life of the C-141, a few examples were selected

During the transition period from European One camouflage to overall gray, paint schemes sometimes got mixed as parts were swapped to keep aircraft mission-ready. A newly-painted gray radome adorns a European One C-141B (63-8081) circa 1992 in the photo.

Shiny and full of promise, the first C-141A (61-2775) was photographed in the Lockheed plant. Its engines are protected by early-style hard inlet covers, tapered to accommodate the inlet cone. (Lockheed Martin)

to receive names, sometimes with christening ceremonies reminiscent of ship launchings. The 1501st Air Transport Wing (ATW) and later the 60th MAW at Travis AFB in California operated the Golden Bear (63-8088), both as a new silver A-model, and later as a C-141B. At McChord AFB near Tacoma, Washington, a new 62nd MAW C-141A was christened *The Tacoma Starlifter*. *The Spirit of Oklahoma City* (C-141A 63-8078) was delivered, with ceremonies and an Air Force band heralding the Starlifter era at Tinker AFB, Oklahoma on October 19, 1964.

When C-141Bs shouldered a large share of Desert Shield airlift duties, quickly stenciled camels on the side of a fuselage could denote a supply mission to bases in Arab countries close to the impending war zone. Less permanent were markings occasionally applied with wide silver tape, sometimes heralding adventures the C-141s engaged in at the annual Airlift Rodeo competition held by the Military Airlift Command.

An expanse of light gray paint makes for a non-standard camouflage coat on this C-141B. (Lockheed Martin)

Hanoi POW Transports

Several gray-and-white C-141As were designated to fly American POWs from Hanoi to freedom at the end of American participation in the war in Southeast Asia. The selected Starlifters were denoted by the addition of a prominent red cross on either side of the vertical fin.

NC-141A 61-2777 retired in September 1994, passing over the open pit borax mine near Edwards AFB on its final flight to the Aircraft Maintenance and Regeneration Center (AMARC) in Arizona. (AFFTC History Office)

The first Starlifter over Georgia shows early markings, a combination of Air Force regulations and test necessities. (Lockheed Martin)

Vestiges of snow cling to a newly painted European One C-141B on the McChord flightline circa 1981. Green tones gave rise to the nickname "Starlizard" when European One colors were first introduced on the C-141. At wet McChord Air Force Base, the camouflaged Starlifters reminded some personnel of a different Northwest creature, and the green C-141 was sometimes referred to as a "Slug."

58 WARBIRD**TECH** SERIES

Two gray C-141Bs in the afternoon sun at McChord AFB in February 1993 carried the green Team McChord tail banner. The dappled appearance of far aircraft is the effect of shadows. When McChord AFB needed to upgrade its infrastructure to accommodate Starlifters, one task was to cut a new opening near the top of the hangar doors (visible behind the large McChord lettering) to allow T-tails to pass into the hangar for servicing.

The Lockheed-Georgia ramp bustled with Starlifter test activity circa 1964. The C-141 nearest the camera (61-2777) had its wing spoilers in various deployed positions when the photo was made. Though C-141 spoilers could be used in flight to promote steep descents, the practice was not favored when passenger comfort was a concern. (Lockheed Martin)

Operation Desert Shield, followed by Desert Storm, demanded high airlift capacity. Some C-141Bs like this 62 MAW example (65-0251) were rushed from depot overhaul before they could be painted. Metal protective coatings gave a gold hue, and rudimentary markings were stenciled in black. Airlifters, always quick with a nickname, soon took to calling these examples "Silver Bullets." Later, when the switch to overall gray paint was impending, a few C-141Bs again flew in bare metal until they could be scheduled for the new coat.

The amber-gold hue of protective coatings on unpainted fuselage extensions of a C-141B graphically shows the placement and size of the forward stretch.

When the overall gray scheme was brand new, this C-141B (65-0277), transferred from the 63rd Airlift Wing, Norton AFB, California to the 62nd Airlift Wing at McChord.

LOCKHEED C-141
STARLIFTER

C-141 Development and Operations

The promise of global reach with air-refuelable C-141Bs led to the designation of some Starlifter aircrews as AD (airdrop) qualified, some as AR (air refueling) qualified, and some qualified to perform both special tasks (AR/AD). In addition to dedicated air refueling training sorties, Starlifter crews honed this skill on special training missions that included various operational scenarios, as well as during real-world missions like the midwinter Antarctic airdrop, which required aerial refueling.

The left petal door of C-141 60208 was lost over Korea during an airdrop mission in March 1984. The Starlifter returned to Yokota Air Base, Japan to await repairs. (63MAW/PA photo by Capt. Greg Smith)

C-141 Development and Operations

The first flight of C-141A 61-2775 on December 17, 1963 began an intensive and integrated flight test program tended by crews and engineers from the Air Force, Lockheed, and the Federal Aviation Administration (FAA). The FAA was closely linked with aircraft numbers one, two, four, and five in the test program. Into its first year of testing, the C-141A was taken as fast as Mach .89 at gross weight. At that time in the program, a speed limitation at Mach .82 was expected to be imposed for normal operations; in service, Mach .767 was the norm (until fuel crises

A Starlifter from McGuire AFB, configured for medevac, airlifted injured American special-operations soldiers from Afghanistan to Ramstein Air Base, Germany, December 7, 2001. Throughout its service life, the C-141 has been readily available to fly the injured to safety. (U.S. Air Force photo by MSgt. Keith Reed)

in the early 1970s gave rise to a more economical cruise setting of Mach .74). It was flown as high as 40,000 feet during its first year.

The first flight lasted 55 minutes, during which the landing gear was kept extended. Some accounts said the flight was made two days ahead of schedule in order to place it on the 60th anniversary of the Wright brothers' original flight. The flight followed three days of taxi runs leading up to the morning of the first flight. Lockheed pilot Leo Sullivan and his four-man crew took the brand new Starlifter to 8,000 feet over north Georgia, testing it in 10-degree banking turns to either side. With the gear down, speeds did not exceed 168 knots.

The curved petal doors proved their ability to withstand flight at 200 knots in the open position, although buffeting initially caused enough concern to warrant a small strake on each door, meeting at the centerline when the doors were in the closed position. This strake did not remain as an operational tool, however. As the C-141A pushed out ahead of any airdrop predecessors in speed, its early open-door forays at 200 knots could not be completed with airdrop tests because parachutes capable of withstanding such speeds were still under development in 1964.[51]

The C-141's original flap extension mechanism proved troublesome when two flap screws were expected to thread together properly on retraction after separating during extension. The screws did not always go together correctly, which resulted in misaligned flaps. A field modification at Edwards Air Force Base in 1964 fixed the problem on the first C-141 by using a non-separating linkage; eventually this became a factory solution.

Using vertical tape gauges on the C-141A's flight deck was not universally hailed at first. These

A loaded airdrop pallet tipped upside down on its fall from a 62 MAW C-141 before righting itself beneath its parachute on its descent to the drop zone. (62 MAW/PA photo by MSgt. Mike Pidding)

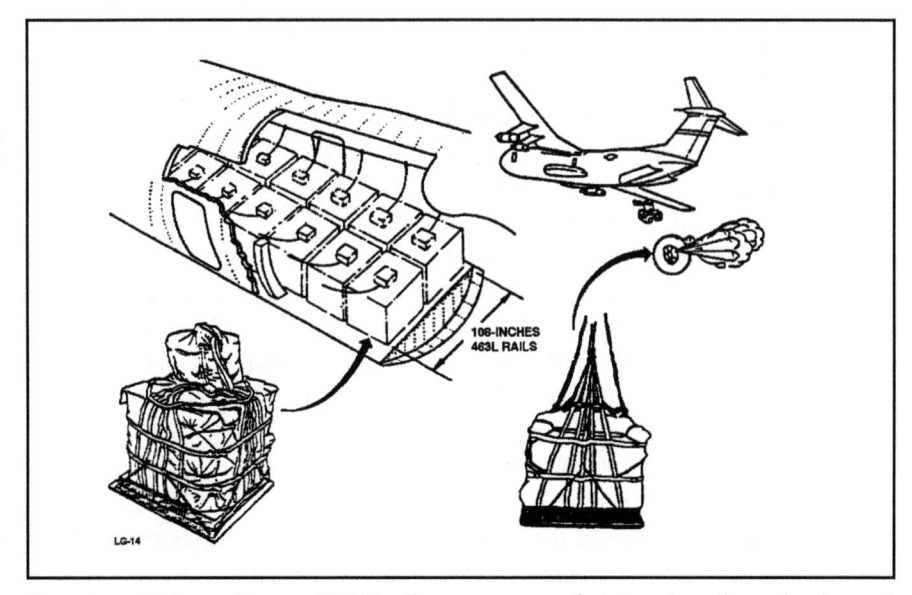

Container Delivery System (CDS) allows two rows of airdrop bundles to be dropped simultaneously. The total time for extracting all the CDS bundles, as well as the spread of the bundles over the ground, can be decreased by increasing the Starlifter's nose-high deck angle and letting gravity assist when the pallet restraints are released. (Lockheed Martin)

This early test Starlifter was photographed over Edwards Air Force Base's flightline at the edge of Rogers Dry Lake, circa 1964. (AFFTC History Office)

Chalk numbers were posted in the pilot's windows of C-141s marshaling at Pope Air Force Base for exercise Gallant Eagle 82 in March 1982. The term "chalk numbers" originated in World War II when numbers would be scrawled in chalk on C-47s to identify which aircraft was to receive which specific load for a mission. For Gallant Eagle, this 62MAW Starlifter was identified as Chalk 43, piloted by Roodzant, and carrying radio call sign Honk 21. Chalk numbers enable mission commanders to ensure that sequenced loads arrive at the right time. In combined heavy equipment and paratroop airdrops, the heavy equipment aircraft drop first, followed by the troops, to ensure equipment doesn't hit any troops who may have difficulty leaving the drop zone.

Paratroops depart a C-141 with jump door wind deflectors extended into the slipstream. After paratroops depart, the loadmasters use a winch to retrieve the parachute static lines from the slipstream. (Photo by Frederick A. Johnsen)

at odds with pilots. The CADC analyzes pressure and temperature inputs affecting the C-141, and comes up with data including altitude, airspeed (true and indicated), and Mach number. The CADC system can also illuminate warning lights when it detects malfunctions. Lockheed made the transition to vertical tape gauges more comfortable for some crews by installing traditional round "steam gauge" airspeed and altitude indicators in early test aircraft for cross-referencing. A T-39 Sabreliner received a vertical tape gauge suite to familiarize MATS crews with the C-141 layout at Randolph Air Force Base, Texas. With time, both the vertical tape gauges and the dual CADC system gained aircrew acceptance.

The wedding of accurate instruments to a CADC was emphasized by the Air Force as a necessity to ensure accurate positioning and situational awareness, beneficial to missions like

instruments provided readings by spooling and unspooling a numbered tape past an indexing mark on the face of the instrument. Though reliable, they were unfamiliar. The tape instruments' reliance on a central air data computer (CADC) instead of direct atmospheric readouts may have further placed them

It was still the Jet Age, on the cusp of the nascent Space Age, when it was popular to compare new machines like the C-141 with vintage aircraft like this early Wright Flyer. As many manufacturers had managed to do before, the Starlifter first flew on December 17 – the anniversary of the Wright brothers' first flight. (Lockheed photo)

Modified to carry avionics test packages, C-141A 61-2777 was photographed on a rainy spring day at Wright-Patterson AFB, Ohio, in 1984. A non-standard outward-hinging crew entry door is evident in the photo. The ASD tail band signifies Aeronautical Systems Division.

airdrops. The CADC, in addition to supporting the flight instruments, also governs the controls' artificial feel system.

Strategic Missions, Airdrop Capability, Minuteman Missile Transport

Loading a Starlifter takes much more preparation than stuffing the kids and the dog in the back of a station wagon for a trip to the beach. MAC developed several standard configurations that could be further modified to create airlift flexibility, as long as basic tenets like safe tie-downs and center-of-gravity (CG) limits were respected. CG occasionally became an issue for C-141 loadmasters if airlift users did not appreciate how crucial it was to maintain the aircraft's balance within a specified range. Cargo weights had to be accurately determined before placing pallets or vehicles in the C-141.

A list of standardized equipment to be carried aboard C-141As provided a baseline of items and uniform placement for them aboard Starlifters. The items and weights on the list were intended to be constant and were included in the listed basic weight of the aircraft. For C-141As, this included everything from one emergency hand axe to nine headsets and microphones. A 1971 MAC listing of equipment called for 100 oxygen masks and hoses to be installed along the sidewalls of the cargo compartment for passengers in the event of a loss of cabin pressure. Airlift crews were mindful of the placement of the masks; if a crewmember ventured farther aft than the available masks, a walk-around oxygen bottle was carried for emergency use. It was not unusual for off-duty crewmembers to seek a place to nap amongst the cargo, cradling an emergency oxygen bottle just in case depressurization occurred. A C-141A table shows the placement of 11 walk-around oxygen bottles, including one in the crew latrine.

Shelves in the crew compartment held technical publications. In the event of an in-flight anomaly, the flight engineer could refer to the publications in consultation with the aircraft commander to determine the proper course of action. At other times, an experienced flight engineer might use the technical manuals during a long mission leg to quiz a junior engineer, bolstering the new engineer's knowledge and confidence levels.

A liquid oxygen converter resided in the right wheel pod; engine covers and various duct plugs were stowed in the aft compartment of the left wheel pod. Emergency aeromedical evacuation equipment, including stanchions for litters, was included in

A loadmaster from the 60th Military Airlift Wing inspected litters rigged inside a specially configured C-141A that returned POWs to Travis AFB, California during Operation Homecoming. The C-141 flights from North Vietnam via Clark Air Base in the Philippines were set up so every ex-POW had both a seat and a litter on the Starlifter. A complete medical staff and a service escort for each POW accompanied every flight. (Air Force photo via Travis Air Museum collection)

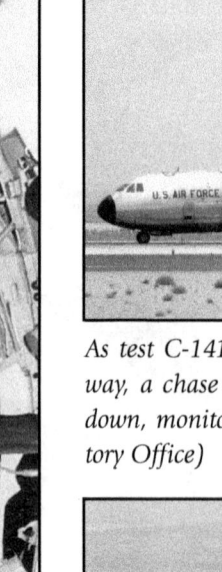

As test C-141A 63-8075 rolls out on the Edwards AFB runway, a chase plane T-38, nose high and with gear and flaps down, monitors the landing from the right side. (AFFTC History Office)

The reflective white paint of the fuselage is preserved in the center wing carry-through, as seen on 62MAW C-141A 65-0229 flying near Mount Rainier. Earlier, this C-141A, then serving with the 60th Wing, made the first landing by a jet aircraft on Antarctica in 1966. (Air Force photo)

the basic equipment. This could facilitate an airlift practice dating back to World War II, enabling the aircraft to carry cargo or troops to a combat area and use the return flight, or back haul, to take casualties away from battle. A C-141 crew might take comfort knowing their aircraft's basic equipment included a pry bar for extending the main landing gear in an emergency, and another bar to release the nose landing gear uplock.

Some things changed over time. C-141As normally carried a sextant, stowed beneath the navigator's table on the left side of the flight deck. When the aerial refueling receptacle and piping atop the cockpit were added, the dorsal sextant port was deleted, rendering the sextant unnecessary.

Letter codes were used to designate different Starlifter transport configurations in paperwork directing missions. Codes included:

AE	–	Aeromedical Evacuation
C	–	Cargo
CP	–	Cargo and Passengers
P	–	Passengers
CAP	–	Combat Airlift (Airdrop) Paratroop
CAC	–	Combat Airlift (Airdrop) Cargo
M	–	Missile
DV	–	Embassy Support Flights, DVs (Distinguished Visitors), etc.
SD	–	Static Display

A Minuteman missile could be transported in a C-141. The Air Force received its first C-141A with Minuteman configuration in January 1966. This fulfilled a national defense need that had been delineated in 1960 as momentum for a fast Starlifter-type transport was beginning to grow. As part of its test program, on April 23, 1966, Lockheed demonstrated a 2.5g symmetrical pullout at Mach .85 with

Continued on page 73

The C-141

Starlizards and Silver Bullets

Glass cockpit instrumentation was part of the upgrades the Hanoi Taxi sported when photographed in 2004.

Starlifter graduates of Lockheed-Georgia returned in the late 1970s and early 1980s to take their place on the shop floor beside new C-130s as Lockheed stretched the C-141As into B-models. (Lockheed Martin)

The famed Hanoi Taxi, C-141A 60177, was photographed at Clark Air Base in the Philippines during its tenure as a POW aircraft, as denoted by the cross on the vertical fin. Decades later, this same historic Starlifter, stretched and upgraded, was earmarked for preservation by the U.S. Air Force Museum. (Doug Remington collection)

The C-141B added luster to the already stellar performance of the Starlifter by making the aircraft aerial refuelable, extending its range and minimizing the need for overseas refueling landing sites. In the photo, the YC-141B tests its refueling ability at Edwards Air Force Base. Wing root leading edge fillets on the YC-141B differ from production fillets. (AFFTC History Office)

A 62MAW C-141A (65-9410), circa 1981, shows the two-tone paint scheme on an active A-model in service. Leading-edge color was sometimes painted gray, sometimes bare metal. The large hose entering the crew door provides hot air for ground crews on cold days.

LOCKHEED C-141 STARLIFTER

Adorned with both the traditional NASA meatball logo on the outboard engine nacelles as well as the since-retired "worm" lettering on the tail, NASA's former civil Starlifter demonstrator became a successful airborne observatory. Photo circa 1994.

With its bright, hopeful civil livery, the sole civil Starlifter flew the skies in search of customers. Lockheed even saw the need for developing ground handling equipment to make airfreight more practical, but the drive-on, drive-off Starlifter was not embraced for quantity civilian production. (Lockheed Martin)

Tail banners recall many of the airlift units that operated C-141s, awaiting the scrapper when photographed in Tucson, Arizona, in February 2005. (Photo by Frederick A. Johnsen)

The non-standard European One C-141B nearest the camera employed darker than normal green paint. Camouflage colors also wrap onto the petal doors of this Starlifter, making door swaps problematical for paint matching. The aircraft were photographed on the 62MAW flightline at McChord AFB, Washington in the 1980s.

Passing through Elmendorf AFB, Alaska in 1965, the C-141A Golden Bear (63-8088) carried the MATS circular emblem behind the fuselage star. A small, and short-lived, strake attached to the petal doors of some C-141As is evident in the photo. The strake, which parted into mirror image halves when the petal doors opened, was installed in an effort to reduce flutter in the petal doors when opened in flight. (Gene Brown via Marty Isham)

The latest all-gray C-141 paint scheme applied to a 445th Airlift Reserve Wing Starlifter as viewed from a tanker in 2000. Insignia and USAF are in black. (Air Force photo by SSgt. Jerry Morrison)

With parted petal doors showing a thin strake that was employed briefly in an effort to reduce door buffeting, C-141A 64-0613 was opened to the public at Lockheed-Georgia's annual Armed Forces Day show, May 15, 1965. (Photo by Kenneth G. Johnsen)

When European One paint was new, some C-141As received camouflage as they came due for painting before stretching. Depicted is 65-9397 of the 63MAW at Norton AFB, California. (Paul Minert via Craig Kaston)

LOCKHEED C-141
STARLIFTER

European One camouflage on a C-141A at Norton AFB, California, on March 26, 1980. Some Air Force painting references said nacelles were to be green (FS 34092), while shaded paint scheme drawings showed nacelles gray, which was typically done. (Paul Minert via Craig Kaston)

Temporary camouflage test on a C-141B at Norton AFB, California, circa March 1982, shows where brown paint was applied to wing flaps before they were extended. (Paul Minert via Craig Kaston)

In the summer of 1982, 63MAW C-141B 65-9404 briefly sported broad irregular panels of desert tan paint for a camouflage test. Earlier tests of tan paint samples over-painted on European One camouflage tested the durability of a temporary camouflage top coat. (Paul Minert via Craig Kaston)

Sometimes called VC-141, this Starlifter, 67-0022, did double duty as a VIP aircraft available to the MAC commander when it was not engaged in training sorties for the 443 MAW at Altus AFB, Oklahoma. This Starlifter was quickly distinguishable from others in the fleet by the white engine cowlings. (Paul Minert via Craig Kaston)

Tugged by static lines, parachutes on CDS bundles begin to unfurl as the packages drop from a C-141B during a demonstration in July 1990.

Test jumpers assigned to the Air Force Flight Test Center parachuted from the YC-141B over Edwards AFB in July 1977. The fuselage stretch and other modifications, including unique wing fillets on the lone YC-141, prompted the need for checking the aircraft's airdrop characteristics. White markings behind the troop door will smudge to reveal any contact with the jumpers or their parachute gear. (AFFTC History Office)

Ghostly gray hues of C-141A 67-0021 of the 438MAW, McGuire AFB, New Jersey, contrasted with brown desert of Nevada during a Red Flag exercise. Camouflage counter-shading is evident in photos of the "gray ghost," which operated for a period in the late 1970s with an experimental paint scheme. Eventually repainted in conventional Starlifter schemes, 67-0021 finally retired as a C-141C in June 2004. (USAF via Craig Kaston)

Chasing the sun that lingers on the horizon during the Antarctic winter, a C-141B cruised during the annual midwinter resupply airdrop to scientific stations. (Air Force photo via Lockheed Martin)

At a time when C-141s were undergoing stretching simultaneously with a move to European One camouflage, the ramp at McChord AFB was a mix of A-models and B-models, and camouflage and white tops, circa 1981.

A pristine C-141A 63-8075 was photographed September 2, 1964 at the Air Force Flight Test Center on Edwards AFB during evaluations. Its mascot and nickname Petunia appear on the nose in front of the crew entry door. Though the MATS band is on the tail, an Air Force Systems Command emblem adorns the nose. "MILITARY AIR TRANSPORT SERVICE" lettering has been applied to the main landing gear pods. (AFFTC History Office)

A C-141B Starlifter of the 62nd Military Airlift Wing banked over western Washington forests and fields near McChord Air Force Base on a murky October afternoon in 1986. A light fixture located on the forward fuselage allowed the crew to inspect the leading edge of the wing and engines in low-light conditions.

LOCKHEED C-141 STARLIFTER

The brilliant afternoon sun illuminated two C-141B Starlifters of the 62nd Military Airlift Wing. This photo was taken at McChord Air Force Base, near Tacoma, Washington in April 1982, with majestic Mount Rainier forming the backdrop. When the Military Airlift Command had a mix of European One camouflaged C-141Bs and gray and white versions, the non-camouflaged aircraft sometimes were selected for overseas missions when diplomatic sensitivities suggested the benefit of a less warlike appearance. (Photo by Frederick A. Johnsen)

In 1976, the year of the American bicentennial, military aircraft with the digits 76 in their tail numbers were often singled out for patriotic paint extras. C-141A 66-0176 featured an insignia leading edge to the vertical fin emblazoned with 13 stars. The official rounded-star American Bicentennial logo also appeared on the vertical fin beneath the Active Duty and Reserve Wing emblems. (Kaston collection)

There's a lot going on in this vintage photo of the C-141A Golden Bear at Travis AFB, California in the 1960s. A non-standard two-tone radome heads the list, followed by the Golden Bear name and the "Lead The Force" decal just to the right. Lead The Force C-141s were deliberately scheduled to fly more hours than most of the fleet to provide indicators of how the Starlifter would age. (Travis Air Force Museum collection)

Continued from page 64

a C-141A in the Minuteman configuration. One of the first five C-141As built flew to Hill AFB, Utah on October 10, 1966, to pick up a Minuteman missile for structural flight testing.[52]

The missile rode in the Starlifter's cargo compartment inside its special Shipping and Storage Container, Ballistic Missile (SSCBM), sometimes referred to by Air Force missileers as the "siscaboom."[53]

The first airlift of an operational Minuteman III missile was made with a C-141A on April 14, 1970. The Starlifter carried the missile from Hill AFB, Utah to Minot AFB, North Dakota.

Backbone of USAF Airlift for Decades

The introduction of the Starlifter to the airlift community in April 1965 gave MATS (later MAC and AMC) the tool with which to redefine itself in mid-Cold War. In addition to milestone firsts, the Starlifter became the reliable, ubiquitous carrier of repatriated hostages, mishap victims, relief supplies, and presidential limousines. The evening news, when focusing on such dramatic events, often framed a glimpse, however brief, of a T-tailed Starlifter, its petal doors open in welcome, as relieved men, women, and sometimes children enplaned or deplaned at the beginning or conclusion of their deliverance.

Airlifters sometimes say they actually perform their missions daily, while other military fliers only practice theirs. It's a boast not meant to demean anyone in the military, but rather to point out the ongoing real-world nature of the airlift business. A U.S. Air Force chronology of the Cold War era, compiled in 1997, gives ample evidence of the standout service the C-141 and its fellow airlifters have provided.[liv] The ability of MAC to respond to natural disasters with aid was documented in November 1970 when C-141As began airlifting personnel and equipment to Dacca, East Pakistan to aid the region as it recovered from devastating tidal waves.

A leisurely cruise over Tacoma, Washington in the mid 1960s by a C-141A visiting from California's Travis AFB was photographed by the Air Force for publicity. C-141s would be a part of the Tacoma airspace for the next three decades. (Air Force photo via Ben Howser)

When testers needed to film C-141A airdrops, the Air Force Flight Test Center at Edwards AFB crafted a series of camera fairings for test airplane 61-2779. In the photo, camera housings are visible on the side of the landing gear pod and beneath it on the belly of the fuselage. (Air Force via Malcolm Gougon collection, Flight Test Historical Museum)

Operation Homecoming

Starting February 12, 1973, specially designated C-141As with red crosses painted on their vertical fins flew into Hanoi to repatriate the first of 590 released U.S. prisoners of war in a scene filled with emotion for all involved. On the first day of Operation Homecoming, 63MAW C-141A 60177 was followed into Hanoi by Starlifters from the 62nd and 60th Military Airlift Wings.

This C-141A with a test boom on its nose made a flaps-extended banking turn for the camera plane with the Lockheed-Georgia factory as a backdrop, circa 1964. (Lockheed via Travis Air Force Museum collection)

One of the last C-141As led a C-130 and the ultimate beneficiary of the Starlifter's jet airlifter trailblazing, the C-17, in a pass during the Edwards Air Force Base air show October 19, 1997.

That year, MAC aircrews were honored by the presentation of the prestigious Mackay Trophy for their Homecoming efforts.[55]

Several MAC Starlifter units participated in this heartwarming mission. As one example, the 62nd MAW from McChord Air Force Base in western Washington logged about 754 hours of flying time in support of Operation Homecoming between February 12 and April 1, 1973. McChord crews flew six missions to Hanoi, four airborne back-up missions, 12 missions from Clark Air Base in the Philippines to homecomings in the United States, and several stateside missions. McChord's 40th Aeromedical Evacuation Squadron, a Reserve unit, contributed flight nurses and medical technicians for some Homecoming sorties.[56]

The Work Never Ends

For a month following October 13, 1973, MAC aircraft including C-141As flew war materiel to Israel during Operation Nickel Grass. From the end of April to mid-September 1975, 251 Starlifter missions augmented 349 commercial aircraft flights to bring a total of 121,562 refugees from Southeast Asia to the United States. Even as that monumental human airlift was taking place in the Pacific, on August 7 two C-141s made a rare Cold War appearance in Bucharest, Romania, to bring disaster relief supplies at the request of the Romanian government in the wake of massive river flooding.[57]

Always the ready tool, C-141s joined C-5s to airlift aircrews and ground personnel to Osan, Kunsan, and Taegu air bases in South Korea on August 20, 1976 after North Koreans killed two U.S. Army officers in an incident that grew from the cutting of a tree. On March 30, 1976, after two commercial 747s collided at Tenerife in the Canary Islands, a joint effort by C-141 and C-130 crews airlifted more than 50 survivors to several bases in the U.S.

When a Soviet satellite fell from orbit and disintegrated over northern Canada in January 1978, a MAC Starlifter flew U.S. Department of Energy specialists and equipment to Edmonton, Alberta to help search for radioactive remains. The next month, 12 burn specialists were airlifted by C-141 to Waverly, Tennessee to treat victims of a propane tank car explosion on a railroad. Hostilities in Zaire's Shaba province prompted 42 USAF C-141 missions and one C-5 mission to carry cargo, fuel, and passengers to Zaire in support of French and Belgian operations in that country in the summer of 1978. The last two months of the year brought diverse real-world C-141 sorties, including the airlift on November 22 of six medical specialists plus their equipment from Boston, Massachusetts to Algiers to tend the critically ill president of Algeria. Six days later, MAC Starlifters carried 911 bodies from the site of a mass suicide at Jonestown, Guyana to the military mortuary at Dover AFB, Delaware. On December 9, 1978, C-141s and C-5s carried about 900 U.S. dependents out of Iran as political tensions increased.[58]

MAC C-141s flew daily in both scheduled and special assignment airlift missions to keep the United States' massive Cold War defense forces supplied globally. Humanitarian and emergency missions were interspersed, as noted in an Air Force chronology for 1979, when C-141s moved 21 tons of relief materials to Fiji in the wake of Typhoon Meli in April; later that month Starlifters flew 20 tons of vegetable seeds to Zaire in an effort to help the population grow food. In October 1979, a pair of C-141s outfitted for aeromedical evacuation transported 38 severely burned Marines from Yokota Air Base, Japan to Kelly Air Force Base, Texas in the aftermath of a fire in an enlisted barracks at a Marine base in Japan. It was a litany that would be repeated over and over for Starlifter crews, responding to the effects of earthquakes, floods, storms and famine around the world.[59]

Special Operations Low Level (SOLL)

A number of MAC C-141 crews trained for SOLL (Special Operations Low Level) flights. In 1994, SOLL II involved 13 C-141Bs fitted with special gear to enhance their capabilities for such missions. SOLL II Starlifters were equipped with a chin-mounted forward-looking infrared (FLIR) sensor as well as two noticeable cheek blisters for dispensing flares and chaff to defeat missile attacks. Some other Starlifters in airlift service received flush-mounted flare dispensers as well

There's a war on as a C-141A rolls along the runway at Da Nang in South Vietnam in 1970. Starlifters quickly proved their worth with rapid long-range missions to and from Southeast Asia beginning in 1966. (22nd Air Force via Travis Air Force Museum collection)

SOLL II Special Operations Low Level C-141Bs featured cheek blisters housing flare and chaff dispensers. Black chin projection is for forward looking infrared (FLIR) device.)MSgt Gary Boyd, 305 AMW/HO)

as infrared detectors to elude missile attacks; these fixtures should not be confused with the bulging devices applied to the SOLL II C-141Bs.

Other late-in-life modifications applied to some Starlifters included a Plexiglas blister to allow a scanner to watch for surface-to-air missile launches.

Antarctic Resupply

During Antarctica's summer months – exactly opposite of summer months in the northern hemisphere – C-141s have made landings at McMurdo Sound on the edge of the continent to resupply researchers there. But during the depths of the

C-141A 65-0243 of the 62MAW flew Operation Homecoming missions to return POWs from Hanoi in 1973. (62MAW/PA)

The ramp is already partly open as 63MAW C-141A 66-0177, marked with a red cross for Operation Homecoming, taxies in. This aircraft made the first flight to Hanoi to pick up freed prisoners of war on February 12, 1973. (62MAW/PA)

This early test photo of C-141A 61-2775 suggests the vertical fin and surrounding upper fuselage structure were painted from the start, with the petal doors natural metal at the time the photo was taken. Petal doors subsequently began acquiring painted sections as well. (Lockheed Martin)

Antarctic winter in June, such landings are ruled out, necessitating supply airdrops instead. Container Delivery System (CDS) boxes can be airdropped over McMurdo as well as over the South Pole station in the middle of the ice continent. When Starlifter crews plotted the distance from the South Pole station back to safe haven in New Zealand, they contemplated the potential for having the C-141's petal doors freeze in the open position. The drag would consume more jet fuel than could be spared for the return to New Zealand, spelling an icy or watery end to such a sortie. For this reason the back ramp and petal doors can be used for the airdrop at coastal McMurdo, but for the ultimate deliveries to the South Pole, crews roll open the two paratroop doors flush with the fuselage and push the CDS bundles out either side. Warehouse style conveyor rollers have been placed on the C-141 floor to facilitate moving the containers to the troop doors for the South Pole drops.

Grim humor accompanies this task in 100-degree-below-zero temperatures: crewmembers in the back of the Starlifter, wearing parachutes because the aircraft is open, have been advised if they tumble out of the aircraft they can either pull the

Airlift fundamentally changed on December 17, 1963, with the first flight of C-141A 61-2775 from the Lockheed plant at Marietta, Georgia. (Lockheed via AFFTC History Office)

On a rain-spattered July 9, 1972, C-141As from McChord AFB dropped troops in an air show demonstration over their home field. Airdrop formations are worked out with lateral and fore-and-aft separation between aircraft designed to preclude collisions between jumpers and aircraft. (Photo by Frederick A. Johnsen)

rip cord and survive the fall, only to freeze in the frigid darkness before help could possibly find them, or they can elect not to open the parachute and shorten the inevitable. Fortunately, nobody has had to make this choice.

The midwinter Antarctic airdrop missions stage out of Christchurch, New Zealand and rely on a KC-10 for an aerial refueling on the way to Antarctica.

Operation Urgent Fury

The Starlifter went to war in October and November 1983 when U.S. troops invaded Grenada in the Caribbean to protect U.S. citizens there in the face of a coup.

Giving Comfort

For decades, C-141s have carried victims of global violence to safety, as on December 11 and 12, 1984, when a pair of Starlifters flew survivors, and two casualties, from a hijacked Kuwaiti airliner to Rhein-Main Air Base, Germany, and the U.S. Usually overlooked during such emotional and newsworthy events, news cameras nonetheless frequently captured

A recently liberated American POW boarded a C-141A for the trip home to the United States in 1973. (62MAW/PA)

the presence of a tall T-tail. On July 1, 1985, Vice President George Bush greeted 39 passengers from a hijacked TWA flight, who were released and subsequently carried by a 438th MAW Starlifter to Rhein-Main from Damascus, Syria. That October 11 and 12, another 438th MAW C-141 mission brought 11 released survivors from the hijacked cruise ship *Achille Lauro* from Cairo, Egypt, to Newark, New Jersey.

If charity begins at home, then C-141 humanitarian missions helped people in the U.S. as well as abroad. During the last half of July 1986, as unusually persistent drought conditions threatened livestock herds in seven southern states, 24 C-141 missions, along with C-130 sorties, brought nearly 536 tons of donated hay to save the livelihoods of many farmers during Operation Southern Haylift.

Cheers and elation were shared by American ex-prisoners of war as one of the specially marked C-141A Starlifters carried them to freedom in early 1973. (Air Force via Travis Air Force Museum collection)

Medical evacuation buses and ambulances could load directly from the ramp of a Starlifter, as shown by C-141A 64-0627. (Travis Air Force Museum collection)

During a number of fire seasons, C-141s delivered fire crews, equipment, and supplies to combat large wildfires in the United States. A major effort was made from August to October 1988 as 13 fires ravaged Yellowstone National Park in Wyoming and Montana.

Just Cause

Five days before Christmas 1989, MAC Starlifters participated in Operation Just Cause, during which 9,500 U.S. troops were flown to Panama and Panamanian leader Manuel Noriega was removed from power.[60]

Starlifters airdropped troops of the Army's 82nd Airborne Division, as well as heavy equipment, in two formations the night of December 20 and 21. During this combat airdrop, a loadmaster on one C-141 from McChord Air Force Base saw ground fire pass between his Starlifter and the aircraft behind it. Although no hits were confirmed, the punctuation by tracers in the night left no doubt this was actual combat for the C-141.[61]

Desert Shield and Desert Storm

America's first war with Iraq in early 1991 began with a prelude in the last part of 1990, during which materiel and troops were prepositioned in the Middle East for combat. The trigger for this activity was the invasion of Kuwait on August 2, 1990, by Iraqi forces. Six days later, the first USAF transport to land in the crisis area was a C-141 flown by a Reserve aircrew.

Statistics for intertheater airlift conducted during the build-up phase, known as Operation Desert Shield, show the C-141 to be the major workhorse of that effort. Again during the combat phase, called Operation Desert Storm, C-141 missions far outstripped other forms of airlift. The first mission of Desert Shield was flown by a C-141 Starlifter on August 8, 1990. During Desert Shield, C-141 intertheater missions totaled 5,549, more than twice the number of the next highest airlifter. These C-141 missions accounted for 30,319 sorties. During the Desert Storm combat phase, Starlifter missions again more than doubled those of the nearest competitor. C-141s carried 72,635 passengers on intertheater Desert Shield missions, and 16,751 passengers on intertheater Desert Storm missions. Figures compiled by the Air Mobility Command history office include:

Desert Shield Intertheater Only, August 1990 – January 15, 1991

Aircraft	Missions	Sorties
C-5	2,648	13,729
C-141	5,549	30,319
C-130	638	2,288
KC-10	378	1,825
Commercial	1,963	7919
Aircraft	Passengers	Tons/Cargo
C-5	68,998	157,608
C-141	72,635	103,461
C-130	5,324	4,123
KC-10	1,112	12,055
Commercial	249,184	87,242

Desert Storm Intertheater Only, January 16 – February 1991

Aircraft	Missions	Sorties
C-5	993	4,808
C-141	2,457	13,016
C-130	320	1,121
KC-10	0	7
Commercial	1,211	4,791
Aircraft	Passengers	Tons/Cargo
C-5	17,098	62,053
C-141	16,751	48,717
C-130	2,031	2,460
KC-10	0	94
Commercial	58,338	52,274

With flaps down and wind deflectors deployed from the troop doors, a C-141B from the 437MAW released paratroops over Eglin AFB, Florida in 1981. This white-top Starlifter carries the gray paint of the wing box structure across the top of the fuselage; some aircraft had the wing box painted white atop the fuselage. (Travis Air Force Museum collection)

Fairings came and went as NC-141As of the Aeronautical Systems Division (ASD) tested avionics for use on other aircraft. The large fairing may be for a Sapphire radome.

If intratheater missions had been tracked, C-130 missions would surely have been higher than just the intertheater statistics indicate. But for the intercontinental strategic airlift, C-141s were the bread and butter.[62]

Into the Sunset

C-141s continued their service around the globe in the 1990s, including emergency relief missions to portions of the former Soviet Union and elsewhere, even as the Starlifter's replacement, the C-17, gained operational capability incrementally and entered the airlift stream. The second war with Iraq in the spring of 2003 put a dwindling supply of Starlifters to work in that region, while other C-141s awaited scrapping at the Aircraft Maintenance and Regeneration Center (AMARC) at Davis-Monthan Air Force Base in Tucson, Arizona. Just as C-124s quickly made way for the new jet Starlifters in the last half of the 1960s,

As the last CDS bundle tipped off the back ramp of C-141B 65-0229 on June 24, 1983, a loadmaster signaled thumbs-up for a successful midwinter resupply drop over McMurdo Sound, Antarctica. Midwinter conditions prevented landing on Antarctica, so C-141Bs made perilous airdrops with open Starlifters in temperatures that reached 100 degrees below zero Fahrenheit.

A palletized Army vehicle dived for earth as it departed a C-141B during the Gallant Eagle mass airdrop exercise in March 1982. A deployed extraction chute pulled the vehicle's pallet from the back of the Starlifter; a massive multi-parachute bundle will open momentarily to bring the vehicle safely to earth on the Fort Irwin reservation in California's Mojave Desert. Exhaust from the C-141B partially blurs it in the photo.

Within performance parameters, it was possible to fly the Starlifter with combinations of wheels, flaps, and petal doors deployed, as photographed circa November 1982.

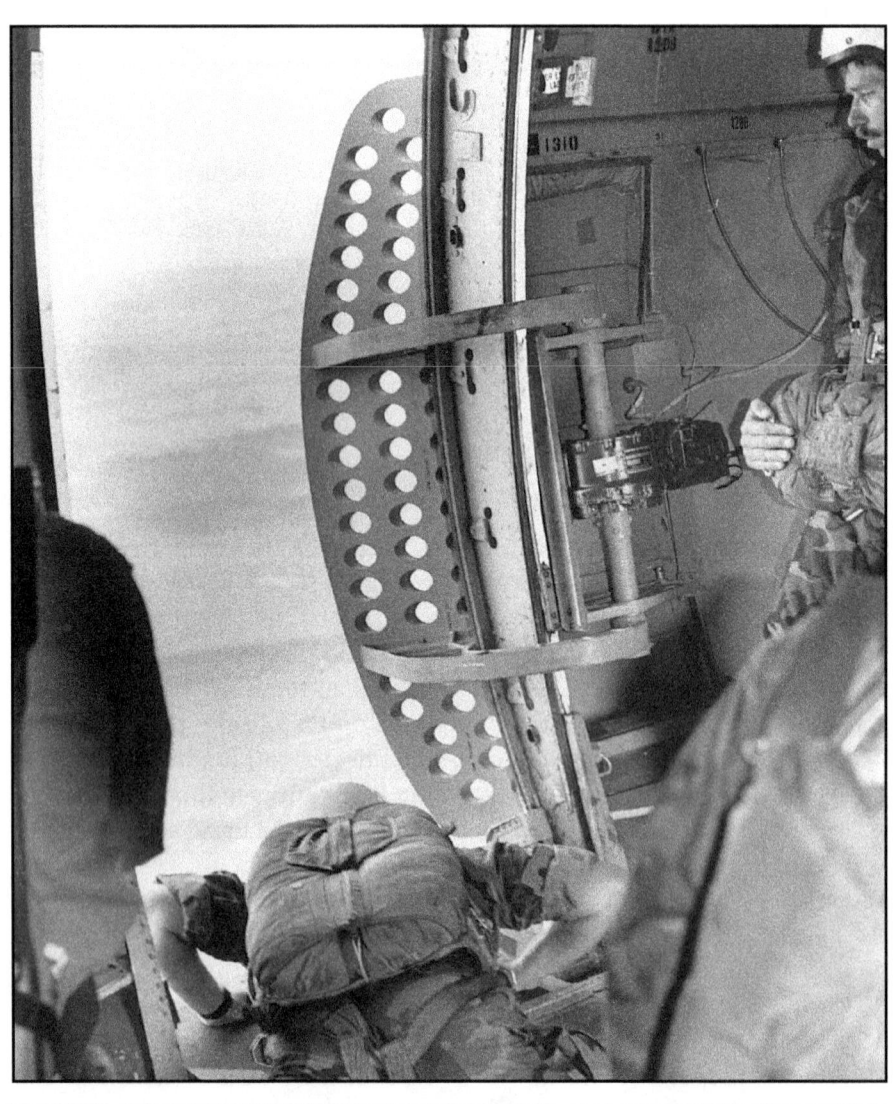

A jumpmaster wearing a parachute peers from the open left troop door of a C-141B nearing a drop zone on Oahu. A perforated wind deflector is extended into the slipstream to facilitate safe departure for the paratroops. Whenever any doors are open in flight, all personnel in the cargo compartment must either wear parachutes or restraining harnesses.

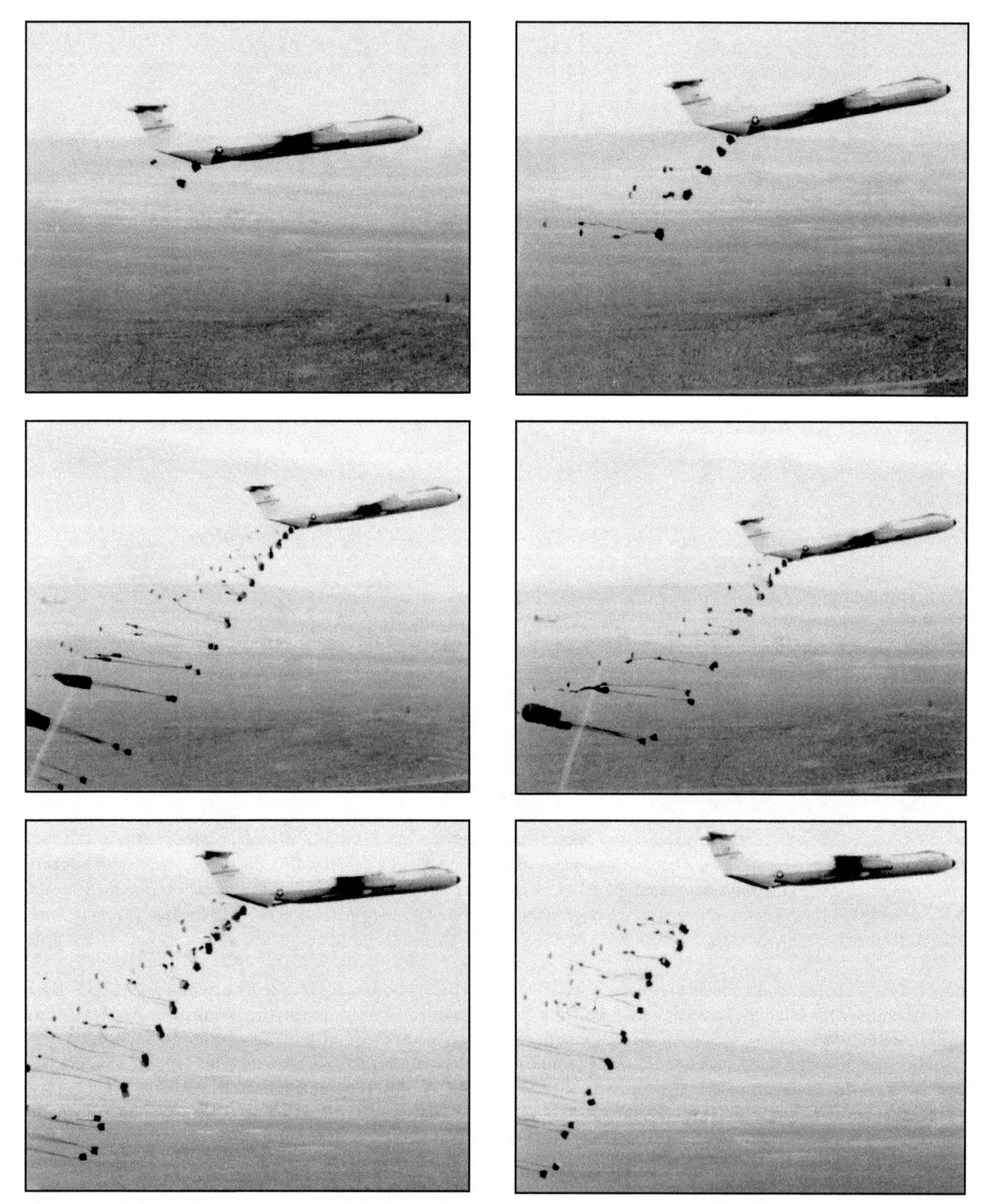

This series of photos documents gravity-aided release of Container Delivery System (CDS) bundles from a test C-141B over Edwards AFB. Dropping with the nose higher resulted in a tighter ground pattern for bundles; drop attitude could be set to aid specific mission parameters. (AFFTC History Office)

Troop parachutes billowed and eventually blossomed as jumpers left a C-141B during a 1989 demonstration. (Photo by Frederick A. Johnsen)

The end of the line for Starlifters was photographed in February 1994 at Davis-Monthan AFB, Tucson, Arizona, where the Aircraft Maintenance and Regeneration Center (AMARC) stores and dismantles military aircraft. (Photo by Tony Accurso)

Broken beyond repair, a C-141 at the Aircraft Maintenance and Regeneration Center (AMARC) in Tucson, Arizona, in February 2005 symbolizes the end of the line for most Starlifters. A large portion of the Starlifter fleet was quickly sent to the scrapper. Some arriving C-141s did not receive protective storage treatments, since they were destined to be scrapped soon. (Photo by Frederick A. Johnsen)

now C-141s were put to pasture as C-17 Globemaster IIIs filled the ranks of the Air Mobility Wings.

The 1994 Scientific Advisory Board convened by the Secretary of the Air Force determined flying Starlifters beyond the equivalent of 45,000 flying hours could jeopardize the design's inherent fail-safe features because of the onset of widespread fatigue.

The first C-141 flew to retirement at AMARC in 1993. By the end of September 2000 (closing out the Air Force's fiscal year), 131 C-141s, less than half the fleet, remained in service. End of service, projected for 2006, loomed large. C-141Bs began moving into the Air Force Reserve and Air National Guard as early as 1986. By 2004, the last nine active-duty C-141Bs were stationed at McGuire Air Force Base, New Jersey. Reserve C-141Cs were at March Air Reserve Base, Riverside, California and Wright-Patterson Air Force Base, Ohio in the first half of 2004; seven Air National Guard C-141Cs flew from the Memphis Air National Guard Base in Tennessee in 2004.

At least one C-141 bore the wounds of combat. On July 21, 1994, a C-141B participating in Operation Provide Promise humanitarian missions over Bosnia encountered small-arms fire near Sarajevo. More than 25 holes were counted in the fuselage and wings after the Starlifter landed at Rhein-Main Air Base, Germany. This led to a temporary suspension of all flights to Sarajevo.

Supporting Iraqi Operations

There are never enough transports when a contingency erupts. Ask any airlifter of any era. By 2003, the C-17 Globemaster III was firmly ensconced in the Air Force's airlift role, replacing Starlifters in the active duty Air Mobility Command wings. Yet Starlifters were still pressed into service to support Operation Iraqi Freedom as American and coalition forces swept over Iraq and Saddam Hussein's government toppled.[63]

Born at the height of the Cold War to provide flexible response capability, the Starlifter has given more than 40 years of adaptable airlift. Whether helping to keep a seemingly monolithic Cold War adversary at bay, or later flying humanitarian supplies to the now dissembled elements of that same former adversary, and subsequently entering the stark new ideological landscape of counter-terrorism activity, Lockheed's C-141 Starlifter and the airlifters who animate it have placed a permanent marker in the advancement of aeronautical capability.

The MAC Starlifter Family

With some evolution of unit designations, the active-duty airlift community based Starlifters at several stateside bases for many years before the advent of C-17s sent bands of C-141s off to new jobs. In 1966 the Military Airlift Command (MAC) replaced the venerable Military Air Transport Service (MATS) as the Air Force's airlift agent. A series of redesignations and activations set up the Military Airlift Wings (MAWs) that would exploit the C-141 around the world. C-141 active-duty Airlift Wings and their home stations, evolved as part of the 1966 stand-up of MAC, were:

A C-141B takes on fuel over the north Pacific with its T-tail in the contrails of the KC-10 refueling it.

C-141As figured prominently in the movement of Apollo astronaut crew quarantine trailers during moon missions. (via Lockheed Martin)

60th MAW, Travis AFB, California
62nd MAW, McChord AFB, Washington
63rd MAW, Norton AFB, California
436th MAW, Dover AFB, Delaware
437th MAW, Charleston AFB, South Carolina
438th MAW, McGuire AFB, New Jersey
443rd MAW (Training), Altus AFB, Oklahoma

Air Force Reserve Wings were associated with active-duty Wings, flying the same C-141s at a number of bases. Using the Starlifters assigned to the Active Duty Wings, the Reserve C-141 workload was substantial; as an exam-

A C-141A dusted off the surface of Rogers Dry Lake at Edwards AFB during testing. (Lockheed Martin)

Facing well-wishers, a POW from Vietnam returned the waves of his greeters in a scene repeated many times in 1973. (Air Force photo)

Propped and chopped, C-141Bs in storage at AMARC in Tucson, Arizona began giving up parts to keep others flying by the mid 1990s. (Don Buescher)

C-141s landed on Grenada in 1983, using a captured runway that had been under construction by Cuban laborers. (Air Force via Lockheed Martin)

The cylindrical test fixture added to the aft fuselage of C-141A 61-2777 necessitated modifications to the petal doors, which nonetheless remained functional for ground loading.

Overall gray C-141B (subsequently upgraded to C-model) 66-0130 flew with the 183rd Airlift Squadron of the 172nd Airlift Group at Jackson, Mississippi when photographed in August 1992. (Al Mongeon via Craig Kaston)

ple, McChord's 446th MAW (Associate) often flew one in three Starlifter missions mounted by the 446th and its Active Duty host, the 62nd MAW. The 63rd MAW was augmented by the Reserve Associate 445th MAW; at Charleston, the 315th MAW shared Starlifters. Travis' 349th MAW (Associate) supported C-141 and C-5 operations of the Active Duty 60th MAW.

As C-141Bs and Cs migrated from Active Duty Wings to the Air National Guard (ANG) and Air Force Reserve units starting in 1986, Starlifters showed up in ANG outfits including the Tennessee ANG's 164th Airlift Wing and the Mississippi ANG's 172nd Airlift Wing (AW). Reserve units owning C-141s during this late era included the 445th AW at Wright-Patterson AFB, Ohio; the 452nd Air Mobility Wing (AMW) at March Air Reserve Base, California; the 459th AW at Andrews AFB, Maryland; and the 732nd Airlift Squadron of the 514th AMW at McGuire AFB, New Jersey. Earlier lineage evolutions included other Reserve units that gained C-141 flying time in conjunction with Active Duty Wings.

Checking a C-141A of the 62nd Military Airlift Wing, Sgt. Ben Howser (right) passed a maintenance binder to another maintainer at McChord AFB, circa 1966. Rivet-head corrosion soon led to the bands of exposed rivets on the fuselage being painted. (Air Force via Ben Howser)

With radomes, windows, and openings sealed with white pliable plastic, C-141Bs congregated in storage at the Aircraft Maintenance and Regeneration Center (AMARC) in Tucson, Arizona in the mid 1990s. (Don Buescher)

Its systems command badge signifies a test flight for C-141A 63-8075 over the Owens Valley north of Edwards AFB. (AFFTC History Office)

Early C-141A 61-2777 served Air Force Systems Command as a systems testbed even before the gray and white paint schemes. (AFFTC History Office)

FLYING THE STARLIFTER

A walk-around C-141B preflight check took on a choreographed appearance at sunrise as two aircrew members readied their Starlifter for flight. Rigorous inspection before each takeoff ensured safe operations.

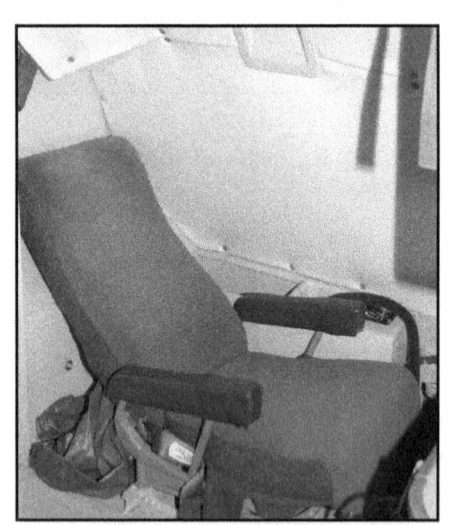

Tucked behind the navigator's station on the left side of the C-141 flight deck is a reclining seat for an additional crew member. C-141 missions often use an augmented crew with more than one pilot, flight engineer, loadmaster, and when needed, navigator, to enable long flights on successive days. (Travis Air Force Museum collection)

With its flexible swept wings and basic control surfaces, the Starlifter was never designed for quick maneuverability, a trait more needed by fighter aircraft. In fact, flight tests and corroborating wind-tunnel testing showed the C-141 could, at high speeds, be vulnerable to aileron reversal. Aileron reversal had manifested itself years before on the slender swept wing of the Boeing B-47 Stratojet bomber. At high speeds, aileron inputs would twist the wing and could cancel out the desired roll input, or, in extreme cases, actually cause the aircraft to roll in the opposite direction from that intended by the pilot. Some C-141 pilots were instructed to avoid high-end Mach numbers because of the potential for encountering aileron reversal, but the practical use of Mach .74 as top cruise speed was below the danger zone.

Description of Handling Characteristics

Maj. Tim Baldwin, an Air Force Reserve C-141 pilot who also flies airline 757s and 767s, says the C-141 is comparable to those airliners (if slower). Perhaps due to the Starlifter's modestly swept wing, it is not difficult to fly. "It's like a big Cessna," Major Baldwin says of the C-141. "It's a very forgiving airplane." When flown lightly loaded for an air show, C-141 crews can climb the big transport to about 10,000 feet in 60 seconds, according to Major Baldwin.

Normally cruised at .74 Mach (sometimes flown at a less fuel-efficient .767 Mach earlier in its career), the C-141 begins to experience a nose-down tuck if flown at .825 Mach, where it becomes less responsive to control inputs and susceptible

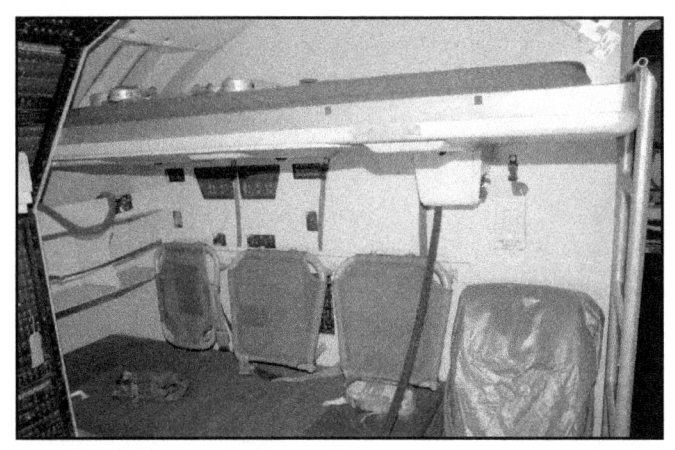

Abutting the aft bulkhead of the C-141 flight deck, upper and lower bunks can allow augmented crewmembers to catch an in-flight nap. Often, the lower bunk served instead as three seats for extra crewmembers; a fourth seat is under a vinyl cover to the right in the photo. It was common for senior flight engineers to take one of these seats while coaching a younger flight engineer seated at the engineer's panel, out of the picture to the left. The ladder for the top bunk, near the right of the photo, also serves as access to an overhead hatch. Aircraft depicted is the Golden Bear, *photographed in 2004 in the Travis Air Force Museum collection. (Travis Air Force Museum)*

The second Starlifter, 61-2776, tested localized aileron control features when photographed in May 1996 during the Electric Starlifter program at Edwards AFB.

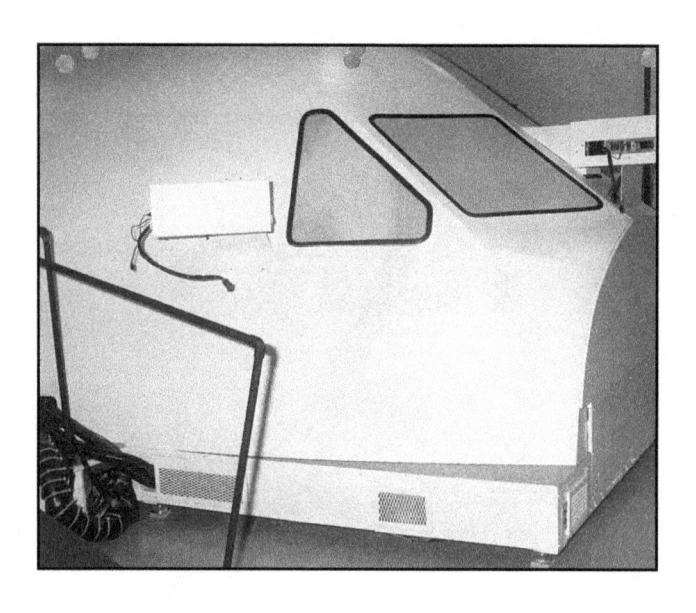

Some Starlifter Wings acquired C-141 cockpit procedures trainers at $1.9 million, like this one pictured at Norton AFB, California. These devices did not use motion like full-up flight simulators, but allowed crews to practice aircrew coordination and emergency procedures. C-141 motion flight simulators were also available. (Travis Air Force Museum collection)

to Dutch roll, a condition where the Starlifter rocks from side to side, indicative that the aircraft is flying faster than its normal operating envelope.[64] Starlifter pilot Keith Littlefield said he always felt the faster .767 Mach cruising speed, pitting the higher thrust against increased drag, made for a more stable flight, requiring less frequent throttle adjustments to maintain the proper speed.[65]

In-flight spoiler deployment to increase descent is only used when the flaps are up, Major Baldwin explains. If flaps are extended, spoiler deployment opens a hole in the wing. A control interlock is designed to prevent accidental spoiler deployment in flight while the flaps are down. Once on the ground, both spoilers and flaps are used to slow the C-141, with the spoilers' ground position nearly perpendicular to the top surface of the wing.

Major Baldwin explains that the C-141's Category II All Weather Landing System (AWLS) can be used in visibility conditions that permit the aircrew just 1,200 feet of forward visibility. The system, with autopilot and auto throttles engaged, is capable of bringing the Starlifter in for a hands-off landing, something the crews practice, but seldom need, on regular missions.[66]

Keith Littlefield, who flew C-141s with the Air Force Reserve's 446th MAW, said he was told during instruction that the AWLS auto-land system intentionally flared the aircraft lower than normal to ensure it would not flare too high above the runway. As a result, when a Starlifter landed itself, "it was abrupt," Littlefield remembered. He would sometimes tell the crew to prepare themselves for an auto land since it could produce a solid jolt upon contact with the runway. "It always felt a little bit abusive to the airplane," he explained. Though he said the system was crude by modern standards, "I always had complete faith" that AWLS would save the day if he ever found himself piloting a C-141 low on fuel and in poor visibility.[67]

Other anecdotal evidence from C-141 aircrews suggests that auto land sometimes induced a flare higher than a pilot would have done. But whether

The gear is tucked away and flaps are set for climbout as a C-141A (65-0235) departs on a flight in June 1979. This view shows why Fowler flaps are sometimes described as "area-increasing flaps." (Photo by Frederick A. Johnsen)

high, low, or on target, Starlifter auto land was a remarkable feature when introduced in the 1960s. Maj. Stu Farmer, who flew C-141Bs with the 62nd Airlift Wing and then early C-141As with the 418th Flight Test Squadron, said a frequently used benefit of auto land was its ability to bring a Starlifter close to the end of the runway in low-visibility conditions that otherwise would have barred landing. Then, with the runway in sight, the pilot could take over at the last minute to finesse the aircraft's landing.[68]

All Weather Landing System Tests

While the Starlifter's purpose-built cargo-hauling airframe was breaking the mold for its class of aircraft, a quiet avionics revolution was taking place in the cockpit as the Air Force, Lockheed, and the FAA worked to achieve all-weather landing capability. In the last two months of 1965, the addition of an automated landing system, for use when visual landings were not possible because of weather, was approved for the C-141. The All Weather Landing System (AWLS) was an ambitious undertaking. By October 6, 1967, the C-141 received FAA certification of AWLS for Category II weather minimums, joining the Boeing 720 jetliner as the only aircraft so certificated at that time. This status meant the C-141 had the ability to land automatically within plus or minus 12 feet left or right of an airport localizer beam, and within 300 feet, plus or minus, of the normal visual touchdown point on the runway.[69]

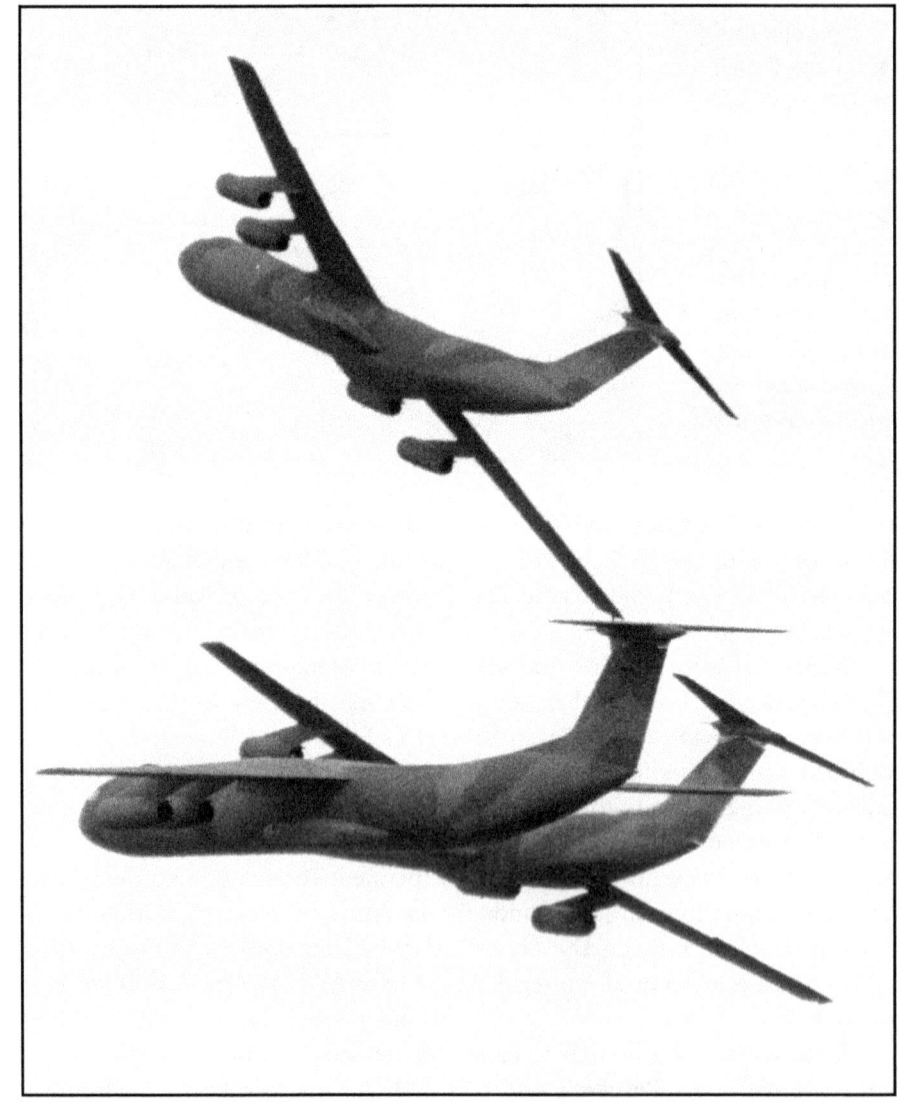

The return of an echelon of C-141Bs to McChord AFB, Washington during a June 1985 air show gave the crews an opportunity to render a four-engine version of a fighter break to landing. (Photo by Frederick A. Johnsen)

According to MAC Starlifter history documentation, adding AWLS reinforced the use of responsive vertical tape flight instruments that came to characterize a few aircraft of the era, including the C-141. These instruments predated the adoption of AWLS in the C-141. AWLS also encompassed auto-throttles, autopilot, a flight director system, radar altimeter, and other components. AWLS was tasked with safely positioning the C-141 for landings under Category II weather minimums, with a 100-foot decision height and a 1,200-foot runway visual range (RVR), as well as demonstrating actual touchdowns under Category II weather criteria. The system was also expected to have provisions to enable the C-141 to align itself out of a crabbed position for rolling out under Category III minimums with a runway visual range (RVR) of only 150 feet, although ultimately the Starlifter's AWLS capability remained Category II. Still, this was no small feat for a 125-ton 1960s-era jet transport hurtling toward contact with the pavement.[70]

The first Starlifter with an interim version of AWLS, aircraft number 66-0132, was accepted by the Air Force on November 21, 1966. Initially, tests of AWLS in the C-141 encountered autopilot malfunctions. Following this, AWLS unit delivery delays meant that the forecast inclusion of AWLS starting on aircraft number 184 was no longer possible; the Air Force's C-141 System Program Office (SPO) directed that production aircraft 157 through 250 would be delivered to the Air Force without AWLS, but with wiring and mounts ready to accept AWLS at a later date. In a preview of problems that would be common in the increasingly computerized Air Force, delays in AWLS demonstrations in 1967 stemmed from interface problems with the logic comput-

Two views capture a 60MAW C-141B (65-0259) departing an en-route stop in September 1983. For a number of years, C-141s of the 60th Wing at Travis were used for strategic airlift missions involving airland, and not airdrop, delivery.

After holding the Starlifter just off the runway in level flight after takeoff to build up speed, the pilot of this McChord C-141B made a spectacular climbout during an airshow performance. (Photo by Frederick A. Johnsen)

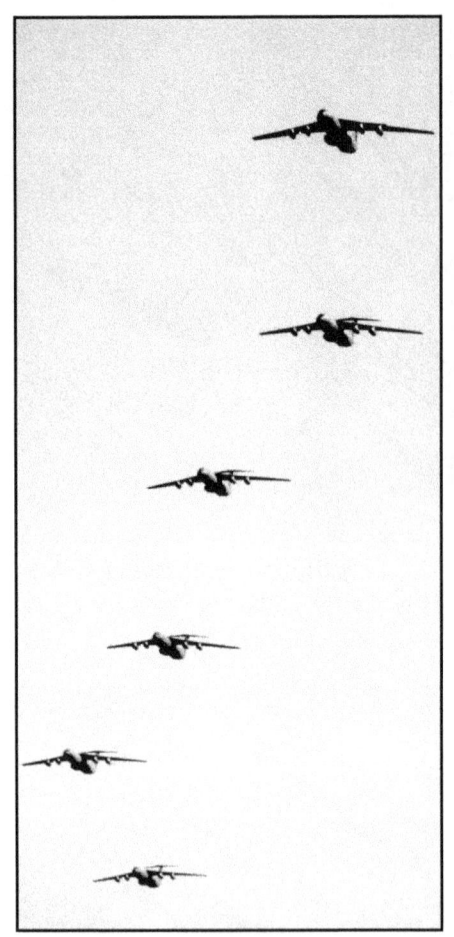

Though often thought of as solitary airlifters, C-141s were frequently used in formations to carry military cargo to a destination en masse. Six C-141Bs lined up in a slightly serpentine echelon for an airfield pass in August of 1991. (Photo by Frederick A. Johnsen)

er, automatic flight control subsystems, and test programmer, all ultimately resolvable.[71]

The October 1967 FAA certification of the C-141's AWLS in Category II circumstances was heralded in the Air Force as a major milestone, giving that service "its first FAA-accepted system which will control the aircraft to touchdown," the Starlifter weapons system manager (WSM) noted.[72] That same month, Lockheed was able to install a complete AWLS suite on the 261st C-141 built; 103 Starlifters delivered with incomplete AWLS gear began receiving the full complement at that time.

But the victory in earning FAA

The first Starlifter, 61-2775, was photographed on a sortie launched on the 30th anniversary of its first flight. When the photo was made on December 17, 1993, the aircraft was part of the 418th Flight Test Squadron at Edwards Air Force Base, California. (AFFTC History Office)

AWLS certification would be hollow unless appropriate bases received instrument landing system upgrades. MAC's request to have 23 bases AWLS-capable by 1971 was overly ambitious in the face of funding cuts. General Estes, MAC commander, evolved a prioritized list of 17 airfields to receive AWLS upgrades: Travis AFB, California; McGuire AFB, New Jersey; Yokota AB, Japan; Mildenhall, England; Elmendorf AFB, Alaska; Dover AFB, Delaware; McChord AFB, Washington; Norton AFB, California; Charleston AFB, South Carolina; Andrews AFB, Maryland; Kelly AFB, Texas; Tinker AFB, Oklahoma; Robins AFB, Georgia; Goose Bay, Labrador, Canada; Hill AFB, Utah; Torrejon AB, Spain; and Wright-Patterson AFB, Ohio. The bases were scheduled for upgrade in batches through 1972.[73]

Pilots needed to be trained on the use of AWLS, and by September 1968, MAC had 100 fliers qualified on the system, even as flightline maintenance personnel were taught how to repair AWLS. By November 1969, even MAC C-141 flight simulators had AWLS capability. But the use of AWLS in C-141s was not without ongoing maintenance and revisions, even as it proved itself in service.

A Starlifter crew availed themselves of AWLS to land at fog-bound McChord AFB, Washington, without incident. The fog was so thick that once the aircraft rolled to a stop, the crew requested to have the aircraft towed to parking because they could not see far enough to taxi the lumbering transport. "They got it on the ground but they couldn't taxi. We had to go out and tow it in," remembered former C-141 crew chief Ben Howser.[74]

Unlikeliest Glider Tug

If line airlift crews have proven their mettle and ingenuity in decades of real-world Starlifter sorties, their arena is still a different

Carrying Lockheed construction number 6100 on its forward fuselage, new C-141A 65-0249 is devoid of most unit markings except the MAC band on the tail in this Lockheed photo. Evident in this view is the lack of a side-facing auxiliary power unit (APU) exhaust in the forward part of the landing gear pod; early C-141A photos show upward-ejecting APU exhaust. (Marty Isham collection)

Deliveries of new C-141As to Air Force units prompted ceremonies like this acknowledging arrival of The Spirit of Oklahoma City *(63-8078). (via Lockheed Martin)*

place than the crucible of the Air Force Flight Test Center (AFFTC) at Edwards AFB, basking in the California Mojave Desert. In the early 1990s, the four remaining NC-141A Starlifters in Air Force service were assigned to the 418th Flight Test Squadron at Edwards.

At Edwards, the number one Starlifter, 61-2775, capped its 35-year career in February 1998 as an unorthodox glider tug, towing an even more unlikely glider – a throttled-back F-106A Delta Dart supersonic delta wing interceptor. The lash-up of Starlifter and Delta Dart, under the Eclipse Project, was a demonstration of a proposed low-

The navigator's station on C-141B Golden Bear, photographed in 2004 after its retirement from service, shows how instruments once used by navigators were blacked out after the advent of computerized components. Many C-141 missions in the post-inertial navigation system (INS) era were flown without an onboard navigator. (Travis Air Force Museum collection)

A view over the pilot's shoulder shows another C-141B in flight during a formation effort over the Pacific Ocean.

cost method of putting payloads into orbit. Kelly Space and Technology (KST) envisioned using larger transports to tow a winged spacecraft to altitude, where it would start its own powerplant for the remainder of the ascent into space. NASA worked with KST on the F-106; the AFFTC furnished NC-141A 61-2775. The effort would be different from most glider tug operations in several ways. Usually, the lightweight, large-spanned glider is airborne before the tow plane leaves the runway; for Eclipse, the C-141 would lift off before the jet fighter behind it gained enough speed to get airborne.

After proposing the demonstration, KST received a small business contract in the summer of 1996 that enabled the project to move ahead. On October 24, 1996, with NC-141A 61-2776 (the "Electric Starlifter," see below) standing in for 61-2775, a

This is how Golden Bear's *copilot station looked in 2004 when it was retired to Travis Air Force Museum. (Travis Air Force Museum collection)*

wake turbulence study was flown in which an F/A-18 fighter, substituting for the F-106, flew untethered behind the C-141 as the fighter's pilot probed for wake turbulence patterns behind the Starlifter.[75]

The Air Force contemplated modifying the Starlifter's ramp with a cutout that would enable the tow line to exit the C-141 while the ramp, petal doors, and pressure bulkhead all remained closed. The final decision was to fly the NC-141A tow plane with its petal doors removed.[76]

Using an over-the-nose bracket, the F-106A was tethered at the end of a 1,000-foot rope consisting of 950 feet of Vectran synthetic woven rope of 3/4-inch diameter and 50 feet of eight-ply nylon rope (for wave damping purposes on the first two flights; after that, 1,000 feet of Vectran sufficed). The rope was only used for one tow flight before replacement. Between December 20, 1997, and February 6, 1998, the unusual tow team made six flight tests. Some tests were conducted at 10,000 feet; the final flight produced data as high as 25,000 feet with the F-106 in tow behind the NC-141A. For the high-altitude tow, the crew in the necessarily unpressurized C-141 was on oxygen and monitored by physiology technicians who flew with them that day.

The F-106 pilot released his aircraft from the tow line to land. The demonstration showed that the idea

*The flight engineer's panel on the right side of the flight deck as it appeared in retired C-141B 63-8088 (*Golden Bear*) in 2004. The red handle in the photo's upper right is an actuator for the refueling receptacle door. (Travis Air Force Museum collection)*

Special noses attached to NC-141A 61-2779 enabled combat aircraft radars to be flight tested. Different noses were used at different times in 779's testbed career. (via Lockheed Martin)

could work, but as of 2004 the ultimate accomplishment of towing a space vehicle and air-launching it into space remains to be realized.[77]

The Electric Starlifter

The second Starlifter built, 61-2776, received a futuristic aileron modification while serving with the 418th Flight Test Squadron at Edwards AFB in 1996. Dubbed the Electric Starlifter (sometimes using the abbreviation ESTAR), the modification placed localized hydraulic systems in the wings to manipulate the airplane's ailerons when activated by electric motors nearby. This replaced the ailerons' reliance on the standard C-141 centralized hydraulic system.

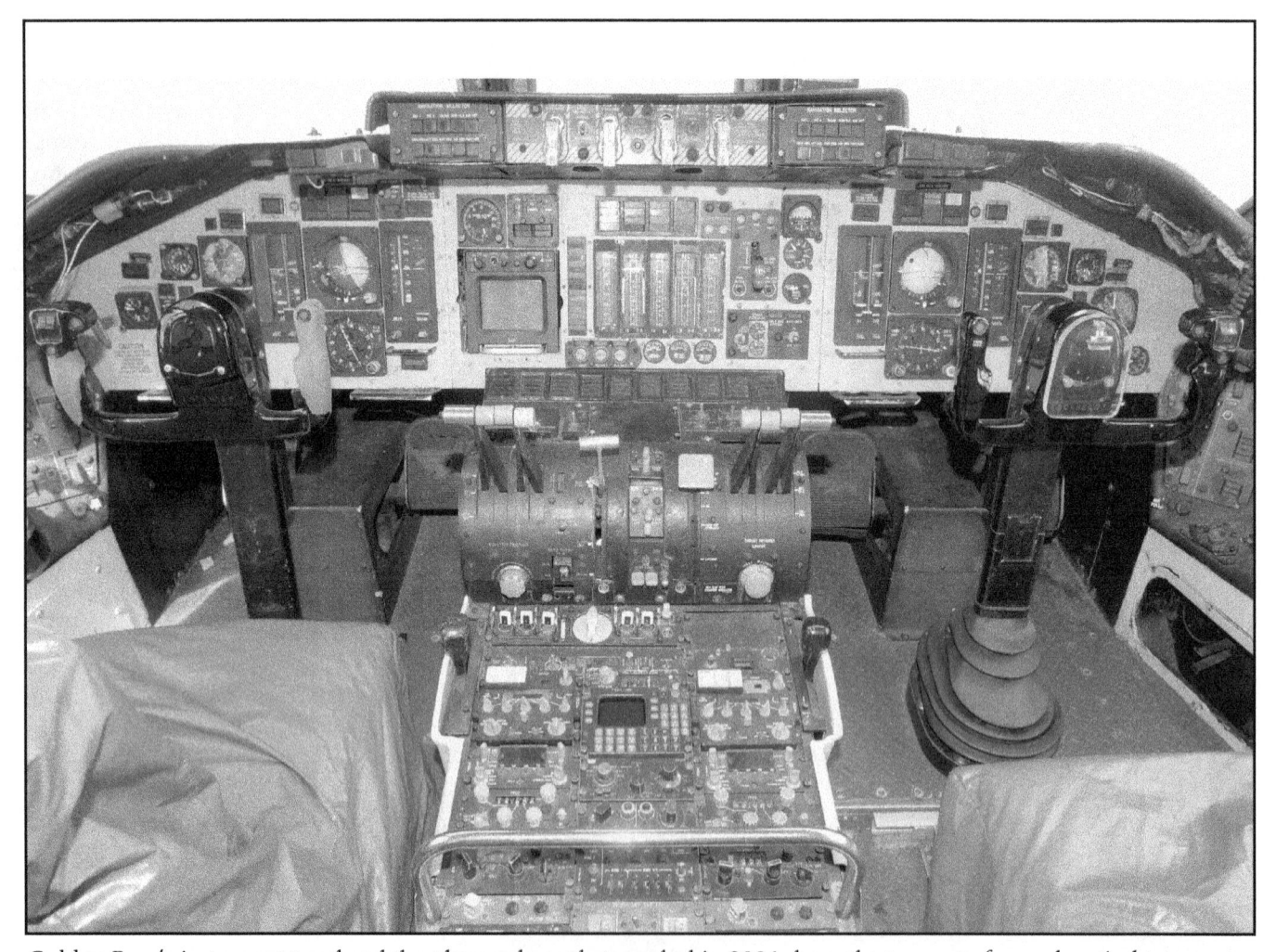

Golden Bear's instrument panel and throttle console as photographed in 2004 shows the presence of several vertical tape gauges on the main panel. The pilots' seats are protected by removable covers. (Travis Air Force Museum collection)

The pilot's side of the flight deck of C-141B 63-8088 shows the nosewheel steering wheel to the left of the control yoke. Pilot and copilot each have a separate set of linked throttle levers on the wide center console. (Travis Air Museum collection)

Savings in weight and serviceability were the intended outcome. The system was for test purposes only; fleet C-141s were not intended to be fitted with such a system, although future warplanes might be.

First flown on April 25, 1996, the Electric Starlifter was exercised to make sure the electrically actuated ailerons did not alter basic C-141A handling characteristics. Following this envelope expansion, the Electric Starlifter logged about 1,000 hours on a variety of airlift missions with a 418th Flight Test Squadron test aircrew in an effort to quantify the new system's reliability and maintainability. The system had redundancies, but even if the entire ESTAR package had failed, the test team computed that adequate roll control for a safe return to base could have been achieved by using trim tabs.[78]

Landing the C-141 was bound to leave a smear of burned rubber on the runway as the mainwheels first contacted the pavement.

A lizard C-141B on its landing roll at McChord AFB in the summer of 1989 used its spoilers to aid in deceleration. The smoke lingering nearby was from some air-show pyrotechnics.

A C-141B rolling to a stop used its spoilers to kill lift over the top of the wing while its thrust reversers swung back to push engine exhaust forward.

A sturdy attach point for a special tow rope enabled this F-106A to be hauled aloft 1,000 feet behind NC-141A 61-2775 for the Eclipse project at Edwards AFB. The Eclipse flights spanned from December 1997 to February 1998; this photo was taken in October 1997. (Photo by Frederick A. Johnsen)

The extensive navigator panel recalls an era when Starlifter missions fully used the navigator's services. In later years, the navigator's equipment dwindled, and some missions were flown without a navigator on board. (Lockheed Martin)

Here's graphic proof that the petal doors were not necessary to ensure structural integrity in flight – the first Starlifter (61-2775) flew with its doors removed to accommodate the towline rigged for Eclipse testing. The first Starlifter finished its 35-year test career in early 1998 with Project Eclipse, towing an F-106 fighter to simulate a space vehicle that could be carried aloft for air launching. (AFFTC History Office)

A white-topped B-model (60129) established its initial climb angle after lifting off from the runway.

Stretched braided line 1000 feet long connects NC-141A 61-2775 with the F-196A it towed for Project Eclipse in 1997 and 1998. Unorthodox lash-up of aircraft with different take-off speeds meant the C-141 became airborne before the F-106 it was towing could lift off, contrary to conventional gliders and tugs. (AFFTC History Office)

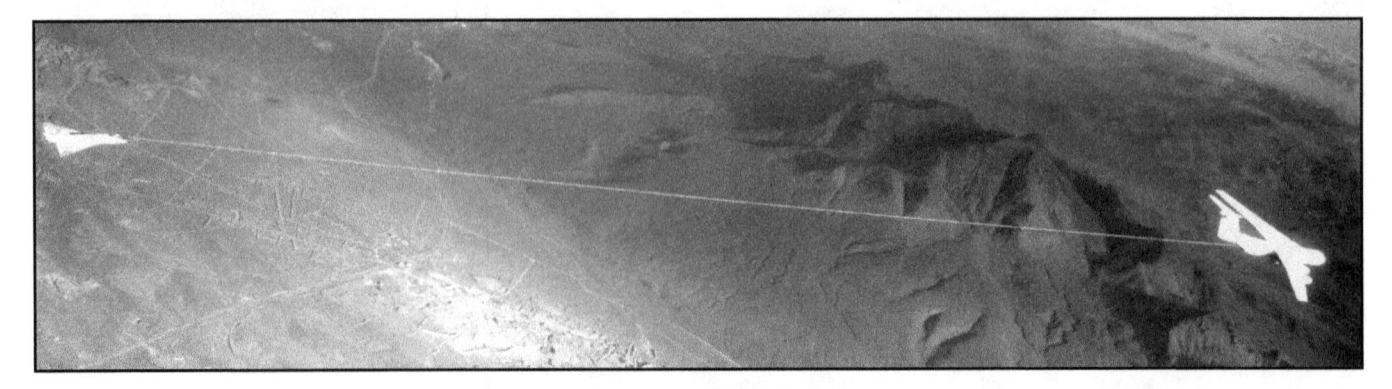

C-141 SERIAL NUMBERS

APPENDIX A

When wing damage made retirement practical for C-141B 64-0642, the fuselage was used for training at Altus AFB, Oklahoma before being turned over to the aviation department of a local college. (Don Buescher)

All 284 Air Force Starlifters were originally delivered as C-141A variants. The lone civil Model 300 Starlifter was essentially built to C-141A dimensions. By 1977, Lockheed won a contract to stretch the C-141As then in Military Airlift Command service, incorporating aerial refueling and other upgrades resulting in the redesignation, without serial number change, to C-141B. In the 1990s, a portion of the C-141B fleet received glass cockpit upgrades, resulting in the nomenclature C-141C.

Four of the first C-141As, after serving the Starlifter developmental flight test program, became airborne testbeds for the Air Force's Aeronautical Systems Division (ASD), ending their service in the 1990s as part of the 418th Flight Test Squadron at Edwards Air Force Base, California. These four Starlifters did not receive B-model stretch upgrades, and they were designated NC-141As to signify that they were not to be used for their original primary purpose of airlift.

Starlifter serials are:
- 61-2775 through 61-2779
- 63-8075 through 63-8090
- 64-0609 through 64-0653
- 65-0216 through 65-0281
- 65-9397 through 65-9414
- 66-0126 through 66-0212
- 66-7944 through 66-7959
- 67-0001 through 67-031
- 67-0164 through 67-0166

C-141C Aircraft Serial Numbers:

As of June 2004, Lockheed Martin counted 63 C-141C aircraft, of which 24 were active and 39 retired. All C-141C aircraft were upgraded from B-models.

Active C-141Cs as of June 2004
- 64-0620
- 64-0637
- 64-0645
- 65-0225
- 65-0226
- 65-0229
- 65-0248
- 65-0249
- 65-0250
- 65-0261
- 65-9412
- 65-9414
- 66-0132
- 66-0151
- 66-0152
- 66-0177
- 66-0193
- 66-7950
- 66-7953
- 66-7954
- 66-7957
- 66-7959
- 67-0031
- 67-0166

Retired C-141Cs as of June 2004

Serial	Date	Note
61-2778	21-Dec-01	Attrited
63-8080	24-Jan-03	
63-8084	31-Mar-04	
63-8085	14-Oct-03	
64-0614	28-Jan-04	
64-0622	28-May-03	
64-0627	4-Mar-04	
64-0632	3-Dec-02	
64-0640	31-Mar-04	
65-0216	6-Oct-03	
65-0222	6-Jan-03	
65-0232	2-Sep-03	
65-0237	31-Mar-04	
65-0245	8-Oct-03	
65-0256	10-Sep-03	
65-0258	14-Oct-03	
65-0271	2-Sep-03	
65-9409	11-Sep-03	

66-0130	25-Feb-03
66-0134	16-Oct-03
66-0136	3-Oct-02
66-0139	16-Dec-02
66-0148	4-Sep-03
66-0157	31-Mar-04
66-0164	19-Feb-04
66-0167	17-Oct-03
66-0174	31-Mar-04
66-0181	31-Mar-04
66-0182	15-Oct-03
66-0185	31-Mar-04
66-0190	20-Nov-03
66-0191	18-Jan-03
66-0201	10-Oct-03
66-7952	6-Oct-03
67-0015	3-Oct-03
67-0021	30-Jun-04
67-0024	31-Mar-03
67-0027	3-Oct-02
67-0029	31-Mar-04

Losses

Though the overall safety record of Starlifters is exemplary, several C-141s have been destroyed in the course of more than 40 years of flying. An Air Force logistics document indicates that 19 C-141s have been lost to accidents and crashes. Losses include:[79]

C-141A 65-0281 (62 MAW) 7 September 1966, McChord AFB, ground explosion.

C-141A 65-9407 (62 MAW) 22 March 1967, ground collision with A-6 at Da Nang, Vietnam.

C-141A 66-0127, (62 MAW) 12 April 1967, crashed on takeoff at Cam Ranh Bay, Vietnam.

C-141A 63-8077 (438 MAW) 28 August 1973, crashed in mountainous terrain east of Torrejon Air Base, Spain.

C-141A 65-0274 (437 MAW) 18 August 1974, crashed in mountains en route to La Paz, Bolivia.

C-141A 64-0641 (62 MAW) 20 March 1975, crashed in Olympic Mountains of Washington state.

C-141A 67-0006 (438 MAW) 28 August 1976, approaching RAF Mildenhall, crashed near Peterborough, UK.

C-141A 67-0008 (438 MAW) 28 August 1976, crashed during landing at Sondrestrom Air Base, Greenland.

C-141 67-0030 (62 MAW crew) 12 November 1980, crashed at Cairo, Egypt on landing.

C-141 67-0017 (438 MAW) 7 March 1982, burned on ground at McEntire Air National Guard Base, Columbia, South Carolina.

C-141B 64-0652 (437 MAW) 31 August 1982, crashed in Smoky Mountains in Tennessee during Special Operations Low Level (SOLL) training flight.

C-141B (437 MAW) 12 July 1984, crashed shortly after takeoff at Sigonella Naval Air Station, Italy.

C-141B (63 MAW) 20 February 1989, crashed on approach to Hurlburt Field, Florida.

C-141B 65-0255 (62 AW) and C-141B 66-0142 (62 AW) 30 November 1992, collided near Harlem, Montana and crashed during night aerial refueling.

C-141B (60 AW) 7 October 1993, explosion and fire on the ramp at Travis AFB, California.

C-141 (438 AW) 23 March 1994, destroyed on the ground at Pope AFB, North Carolina, from fire due to the crash of an F-16 that collided with a C-130 overhead. (Another C-141 also sustained damage.)

C-141 65-9405 (305 AMW) 13 September 1997, collided with a Tupolev Tu-154 off the coast of Namibia.

C-141C 61-2778 (164 AW), 21 December 2001, sustained major wing damage on ground at Memphis, Tennessee.[80]

C-141B 64-0642 and C-141B 66-0154, both grounded for wing cracks, were used as fuselage load trainers at Altus AFB, Oklahoma. The fuselage of Starlifter 0154 later was sold for salvage; C-141 0642 was transferred to a college aviation department at the Altus municipal airport.[81]

Starlifter Accomplishments

C-141A 65-0229 (60 MAW), first landing by a jet aircraft on Antarctica, 14 November 1966.

C-141A 66-0127 (62 MAW) set speed record from McChord AFB to Kimpo, Korea, and back in December 1966.

C-141A 66-0141 (62 MAW), first MAC aircraft to land in Peoples Republic of China, 1 February 1972 in support of President Nixon's visit.

C-141A 66-0177 (63 MAW), first C-141 to bring POWs home from Hanoi, 12 February 1973.

STARLIFTERS B PRESERVED

APPENDIX B

The brand-new Tacoma Starlifter, C-141A 65-0277, was greeted with ceremonies at McChord Air Force Base on August 9, 1966. This namesake aircraft, now a stretched B-model, was retained for the McChord Air Museum as Starlifters were replaced by C-17s at McChord. (Air Force photo via Ben Howser)

Celebrated as the first C-141A in Hanoi to fly American POWs home, C-141C number 0177 was subsequently stretched, upgraded to a C-model, and coated gray overall by the mid 1990s. After this photo was taken, this storied Starlifter was repainted gray and white.

Flying a POW-MIA flag from the overhead hatch in July 2004 at EAA AirVenture in Oshkosh, Wisconsin is C-141C 66-0177, the first Starlifter to take POWs from Hanoi. This C-141 was earmarked for the Air Force Museum upon its retirement.

NC-141A 61-2775, Air Mobility Command Museum, Dover AFB, Delaware. (First Starlifter built.)

NC-141A 61-2779, Air Force Flight Test Center Museum, Edwards AFB, California.

C-141B 63-8079, Charleston AFB, South Carolina.

LOCKHEED C-141
STARLIFTER

Before retirement to the Flight Test Historical Museum at Edwards AFB, NC-141A 61-2779 flew over the Sierra Nevada mountains on a sortie for the 418th Flight Test Squadron in the 1990s.

C-141B 63-8088, Travis Air Force Museum, Travis AFB, California. (*Golden Bear*)

C-141B 64-0626, Air Mobility Command Museum, Dover AFB, Delaware.

C-141B 65-0236, Scott AFB, Illinois.

C-141B 65-0257, March Field Museum, Riverside, California.

C-141B 65-0277, McChord Air Museum, McChord AFB, Tacoma, Washington. (Former *Tacoma Starlifter*)

C-141B 65-9400, Altus AFB, Oklahoma.

C-141B 66-0177, Earmarked for U.S. Air Force Museum, Wright-Patterson AFB, Ohio. (*Hanoi Taxi*, first C-141 into Hanoi to fly POWs home.)

C-141B 66-0180, Robins AFB, Georgia.

C-141B 66-7947, McGuire AFB, New Jersey.

C-141B 67-0013, Pima Air and Space Muesum, Tucson, Arizona.

Last landing for the first Starlifter came on February 22, 1998, as C-141A 61-2775 flew nonstop from Edwards AFB, California, to the Air Mobility Command Museum at Dover AFB, Delaware. (Harry Heist, Air Mobility Command Museum)

Travis Air Force Museum at Travis Air Force Base, Fairfield, California preserved the C-141 christened Golden Bear (63-8088). (Photo via Travis Air Force Museum)

C-141 SIGNIFICANT DATES

APPENDIX C

Making history on November 14, 1966, as the first all-jet aircraft to land on Antarctica, was 60MAW C-141A 65-0229. In 1983, when Starlifter 229 was a B-model serving the 62MAW, it was selected to perform the mid-winter Antarctic airdrops to McMurdo and the South Pole during a time of year when landings are impossible. (The all-jet qualifier for this first landing may take into account the Antarctic use of Navy P-2 Neptunes using jets to augment piston engines.) (U.S. Navy photo by PHC(AC) John D. Reimer, via Travis Air Force Museum collection)

20 December 1960 – USAF issues call for bids to design and build Logistics Transport System 476L, to become the C-141. A month later, Lockheed-Georgia Co. submits its proposal, based on research conducted since 1957.

13 March 1961 – White House announces Lockheed's C-141 as winner of competition to build a fanjet airlifter for the Air Force.

7 April 1961 – Air Force Systems Command (AFSC) executes initial letter contract authorizing Lockheed to begin C-141 program.

September 1961 – First full-scale C-141 engineering mock-up finished; first master model completed. (Master model is a basic contour control tool for gauging fabrication and assembly tooling.)

January 1962 – USAF, FAA and airline teams inspect full-scale C-141 fuselage mock-up; USAF's C-141 Systems Program Office (SPO) formally approves Model Specification.

May 1962 – FAA holds Preliminary Type Certification Board meeting. C-141 will be made eligible for FAA certification, and Lockheed hopes to market the design to commercial cargo carriers.

19 June 1962 – First C-141 assembly, a cargo floor bulkhead, finished.

June 1962 – First TF33 engine and prototype pylon received for use on engine test stand.

September 1962 – First subcontractor C-141 assemblies (cargo floor plates from Bell Aerosystems in Buffalo, New York) arrive at Lockheed.

January 1963 – Required 150-hour engine runs completed by Pratt and Whitney on TF33 test engine for C-141.

March 1963 – C-141 fuselage pressure tests satisfactorily completed.

April 1963 – C-141 wing built by Avco in Nashville, Tennessee joined to first fuselage at Marietta, Georgia. Special railcars transport wings from Tennessee to Georgia.

June-July 1963 – Empennage built by Convair installed on first C-141; four TF33 engines mounted to aircraft.

22 August 1963 – Rollout of first C-141A at Lockheed-Georgia.

17 December 1963 – First flight of C-141, on 60th anniversary of Wright brothers' first flight. Lockheed-Georgia's chief engineering test pilot, Leo Sullivan, makes this first 55-minute flight from Dobbins AFB, Georgia.

1 April 1964 – FAA issues Type Inspection Authorization, pointing toward certification of the C-141's commercial Model 300 variant.

15 June 1964 – C-141 makes first transcontinental flight for delivery from Marietta, Georgia to Edwards AFB, California, where 1,800-hour test program begins two hours after Starlifter's arrival.

23 April 1965 – First operational C-141A delivered to Travis AFB, California.

14 November 1966 – A MAC C-141A from Travis AFB's 86th MAS is the first jet aircraft to land on Antarctica, arriving at McMurdo Sound after flying 2,200 miles from Christchurch, New Zealand.

9 November 1967 – Start of C-141A support missions for the Apollo space program.

17 November 1967 – Start of Operation Eagle Thrust, the longest-distance and largest airlift of troops and cargo from the U.S. to Southeast Asia, involves C-141s and C-130s. By the time the operation is completed, more than 10,350 troops and more than 5,100 tons of equipment reach the combat zone in record time.

23 January 1968 – C-141s participate in troop build-up in South Korea in response to North Korea's seizure of the U.S.S. Pueblo.

28 February 1968 – Last of 284 C-141As bought by the Air Force is delivered to Tinker AFB, Oklahoma.

8 July 1969 – Implementing the U.S. government's Vietnamization policy, the first of 25,000 U.S. troops leave Southeast Asia, boarding C-141As to for the trip from South Vietnam to McChord AFB, Washington.

14 April 1970 – C-141 makes the first airlift of an operational Minuteman III missile, from Hill AFB, Utah, to Minot AFB, North Dakota.

12 February 1973 – MAC crews begin Operation Homecoming, the return of a total of 590 released American POWs from North Vietnam. During this operation, several C-141As receive special Red Cross markings on their vertical fins to identify them as participants in this intense and high-visibility diplomatic undertaking.

21 April 1975 – First military plane to arrive from Saigon carrying non-essential U.S. personnel and Vietnamese civilians following President Gerald Ford's order to step up the relocation of these categories of people was a C-141A landing at Travis AFB. (A World Airways DC-8 carrying passengers from Vietnam arrived in the U.S. earlier that day.)

May 1975 – Contract awarded to Lockheed Georgia Co. to convert a C-141A as the prototype stretched YC-141B.

8 January 1977 – YC-141B, the first Starlifter lengthened by 23.3 feet, rolls out of Lockheed Georgia plant.

25 March 1977 – First flight of YC-141B, at Dobbins AFB, Georgia.

6 June 1977 – The first stretched Starlifter (YC-141B) flies to the Air Force Flight Test Center at Edwards Air Force Base, California.

30 September 1977 – First C-141 transatlantic mission without a navigator is made using a Delco inertial navigation system (INS), between Charleston AFB, South Carolina, and Rota Naval Station, Spain.

December 1979 – First regular C-141B delivered to the Air Force.

6 April 1980 – First operational mission by a stretched C-141B; crew from 443rd MAW flies it nonstop from Beale AFB, California, to RAF Mildenhall in the United Kingdom in 11 hours and 12 minutes, using one aerial refueling.

15 October 1980 – The Golden Bear, first C-141A delivered to an Air Force operational airlift unit in 1965, completes its metamorphosis to become a stretched B-model.

29 June 1982 – Final stretched C-141B rolled out at Lockheed-Georgia.

14 December 1989 – MAC authorizes women to serve as crewmembers on C-141 and C-130 airdrop missions, starting the entry of women into USAF combat crew roles. At least one female C-141 crewmember is said to have flown in the Grenada military operation of 1983, years before the official MAC authorization.

20 December 1989 – During Operation Just Cause, MAC units, including aircraft and crews from C-141 wings, transported 9,500 troops from Pope Air Force Base, North Carolina, to Panama in a time period of less than 36 hours, making this the biggest night combat airdrop since Normandy in June 1944.

October 1997 – Deployment of C-141C models begins.

Mid-2004 – C-141s continue airlift and aeromedical evacuation missions in conflict areas including Iraq.

May 2006 – Final flying Starlifter (66-0177) retired by 445th Airlift Wing in ceremonies at Wright-Patterson AFB, Ohio.

REFERENCES

[1] Frederick J. Shaw Jr. and Timothy Warnock, *The Cold War and Beyond – Chronology of the United States Air Force, 1947-1997*, Air Force History and Museums Program, and Air University Press, Washington, D.C., 1997.

[2] Frederick A. Johnsen, *62nd Military Airlift Wing – Making the History of Air Mobility, 1940-1991*, 62nd Airlift Wing History Office, McChord AFB, Washington, 1992.

[3] Roger D. Launius and Betty R. Kennedy, "A Revolution in Air Transport – Acquiring the C-141 Starlifter," *Airpower Journal*, Fall 1991, Pp. 68-83.

[4] Ibid.

[5] Gen. Howell M. Estes, Jr., "Modern Combat Airlift," *Air University Review 20*, September-October 1969, Pp. 12-25.

[6] From narrative and documents cited in *C-141 Starlifter (January 1959-June 1971)*, Office of MAC History, Military Airlift Command, United States Air Force, Scott Air Force Base, Illinois, January 15, 1973.

[7] Ibid.

[8] Ibid.

[9] Brochure, "U.S. Air Force C-141A StarLifter", Lockheed-Georgia Co., May 1, 1964.

[10] James R. Ashlock, "Air Force Minimized Engineering Changes on Starlifter", *Aviation Week and Space Technology*, September 2, 1963.

[11] Ibid.

[12] Interview, author with Ben Howser, former 62 MAW C-141 crew chief, June 13, 2004.

[13] James R. Ashlock, "Air Force Minimized Engineering Changes on Starlifter", *Aviation Week and Space Technology*, September 2, 1963.

[14] Ibid.

[15] Interview, author with Ben Howser, former 62 MAW C-141 crew chief, April 10, 2004.

[16] See also *Aviation Week and Space Technology*, August 3, and 24, 1964.

[17] *Anything, Anywhere, Anytime: An Illustrated History of the Military Airlift Command, 1941-1991*, MAC Office of History, Scott AFB, Illinois, 1991.

[18] From narrative and documents cited in *C-141 Starlifter (January 1959-June 1971)*, Office of MAC History, Military Airlift Command, United States Air Force, Scott Air Force Base, Illinois, January 15, 1973.

[19] Brochure, *Lockheed C-141: Versatility*, undated, circa 1965, Lockheed-Georgia Co.

[20] William Head, *Reworking the Workhorse – the C-141B Stretch Modification Program*, Office of History, WR-ALC/HO, Robins AFB, Georgia, 1984.

[21] Ibid.

[22] Ibid.

[23] Ibid.

[24] Ibid.

[25] Ibid.

[26] Ibid.

[27] J.R. Wilson, "Power and Glass", *Military Aerospace and NMD Technology*, Vol. 3, Issue 1, 2004.

[28] From narrative and documents cited in *C-141 Starlifter (January 1959-June 1971)*, Office of MAC History, Military Airlift Command, United States Air Force, Scott Air Force Base, Illinois, January 15, 1973.

[29] From interview with Col. William H. Spillers, cited in *C-141 Starlifter (January 1959-June 1971)*, Office of MAC History, Military Airlift Command, United States Air Force, Scott Air Force Base, Illinois, January 15, 1973.

[30] See also *Jane's All the World's Aircraft*, various annual editions, between 1963-87.

[31] Robert H. Cook, "C-141 Completes Successful First Flight," *Aviation Week and Space Technology*, December 23, 1963.

[32] From narrative and documents cited in *C-141 Starlifter (January 1959-June 1971)*, Office of MAC History, Military Airlift Command, United States Air Force, Scott Air Force Base, Illinois, January 15, 1973.

[33] James R. Ashlock, "C-141 Flight Tests Continue on Schedule," *Aviation Week and Space Technology*, August 31, 1964.

[34] David H. Hoffman, "Flow of C-141 Subcontracting Under Way," *Aviation Week and Space Technology*, reprint from the issue of October 9, 1961.

[35] Ibid.

[36] Brochure, *C-141A Starlifter Progress Report No. 7, Featuring Loadability*, Lockheed-Georgia, Marietta, Georgia, May 28, 1964.

[37] "FAA Certificates Commercial C-141: Build Cargo Business Now With Starlifter – Halaby," *Lockheed Southern Star*, Vol. 15, No. 3, February 4, 1965.

38 "C-141 'observatory' discovers atmosphere on Pluto", *Lockheed Star*, June 30, 1988.

39 Brochure, *Lockheed C-141: Versatility*, undated, circa 1965, Lockheed-Georgia Co.

40 Ibid.

41 Cecil Brownlow, "Paris Air Show Stresses Surge by French Aerospace Industry", *Aviation Week and Space Technology*, June 10, 1963, and Herbert J. Coleman, "Argument Continues Over Belfast Future", *Aviation Week and Space Technology*, September 9, 1963.

42 Interview, author with Keith Littlefield, former C-141 pilot, August 15, 2004.

43 Ibid.

44 From narrative and documents cited in *C-141 Starlifter (January 1959-June 1971)*, Office of MAC History, Military Airlift Command, United States Air Force, Scott Air Force Base, Illinois, January 15, 1973.

45 "Our C-141 StarLifter Must Be Made Lean and Strong," *Lockheed Southern Star*, January 4, 1963.

46 Interview, author with Ben Howser, former C-141 crew chief, April 10, 2004.

47 Interview, author with Tim Louden, former C-141 crew chief, April 14, 2004.

48 Interview, author with Ben Howser, former C-141 crew chief, June 13, 2004.

49 From narrative and documents cited in *C-141 Starlifter (January 1959-June 1971)*, Office of MAC History, Military Airlift Command, United States Air Force, Scott Air Force Base, Illinois, January 15, 1973.

50 Interview, author with Ben Howser, former C-141 crew chief, March 2004.

51 James R. Ashlock, "C-141 Flight Tests Continue on Schedule", *Aviation Week and Space Technology*, August 31, 1964.

52 *C-141 Starlifter (January 1959-June 1971)*, Office of MAC History, Military Airlift Command, United States Air Force, Scott Air Force Base, Illinois, January 15, 1973.

53 E-mail, Subj: Re: Query from book researcher – C-141 modified to transport MM? from Charlie Simpson, Executive Director, Association of Air Force Missileers to Ranney Adams, 3 Mar 04.

54 Frederick J. Shaw Jr. and Timothy Warnock, *The Cold War and Beyond – Chronology of the United States Air Force, 1947-1997*, Air Force History and Museums Program in association with Air University Press, Washington, D.C. and Maxwell AFB, Alabama, 1997.

55 Ibid.

56 Frederick A. Johnsen, *62nd Military Airlift Wing – Making the History of Air Mobility, 1940-1991*, 62nd Airlift Wing history office, McChord AFB, Washington, 1992.

57 Frederick J. Shaw Jr. and Timothy Warnock, *The Cold War and Beyond – Chronology of the United States Air Force, 1947-1997*, Air Force History and Museums Program in association with Air University Press, Washington, D.C. and Maxwell AFB, Alabama, 1997.

58 Ibid.

59 Ibid.

60 Ibid.

61 Frederick A. Johnsen, *62nd Military Airlift Wing – Making the History of Air Mobility, 1940-1991*, 62nd Airlift Wing history office, McChord AFB, Washington, 1992.

62 Point Paper, Subj: *MAC Airlift in Desert Shield and Desert Storm by Aircraft*, Action Office: Dr. Kent Beck, HQ AMC/HO, circa March 1994.

63 Walter J. Boyne, *Operation Iraqi Freedom – What Went Right, What Went Wrong, and Why*, Forge Books, New York, New York, 2003.

64 Interview, author with Maj. Tim Baldwin, USAFR, July 2004.

65 Interview, author with Keith Littlefield, August 15, 2004.

66 Interview, author with Maj. Tim Baldwin, USAFR, July 2004.

67 Interview, author with Keith Littlefield, August 15, 2004.

68 Interview, author with Maj. Stu Farmer, USAF Test Pilots School, August 17, 2004.

69 From narrative and documents cited in *C-141 Starlifter (January 1959-June 1971)*, Office of MAC History, Military Airlift Command, United States Air Force, Scott Air Force Base, Illinois, January 15, 1973.

70 Ibid.

71 Ibid.

72 Ibid.

73 Ibid.

74 Interview, author with Ben Howser, former C-141 crew chief, April 10, 2004.

75 History of the Air Force Flight Test Center, Vol. 1, Narrative, 1 October 1995 – 30 September 1997, AFFTC History Office, Edwards Air Force Base, California.

76 Ibid.

77 Tom Tucker, *The Eclipse Project*, NASA Monographs in Aerospace History #23, NASA SP-2000-4523, NASA History Division, Office of Policy and Plans, NASA Headquarters, Washington, D.C., 2000.

78 History of the Air Force Flight Test Center, Vol. 1, Narrative, 1 October 1995 – 30 September 1997, AFFTC History Office, Edwards Air Force Base, California.

79 E-mail compilation of C-141 losses, from Betty Kennedy, AMC/HO, to author, July 7, 2004.

80 From list of C-141C serial numbers, Lockheed Martin, June 2004.

81 From notes furnished by Don Buescher, former C-141 flight engineer.